DYNAMICS OF ARC MIGRATION AND AMALGAMATION - ARCHITECTURAL EXAMPLES FROM THE NW PACIFIC MARGIN

Authored by **Yasuto Itoh, Osamu Takano, Reishi Takashima, Hiroshi Nishi** and **Takeyoshi Yoshida**

Dynamics of Arc Migration and Amalgamation - Architectural Examples from the NW Pacific Margin http://dx.doi.org/10.5772/64579
Authored by Yasuto Itoh, Osamu Takano, Reishi Takashima, Hiroshi Nishi and Takeyoshi Yoshida

Scientific Reviewer:

Mohammad Mokhtari

Contributors

Yasuto Itoh, Osamu Takano, Reishi Takashima, Hiroshi Nishi and Takeyoshi Yoshida

Published by InTech
Janeza Trdine 9, 51000 Rijeka, Croatia

Publishing Process Manager Ana Simcic
Technical Editor SPi Global
Cover InTech Design team

First published May, 2017
Printed in Croatia
Legal deposit, Croatia: National and University Library in Zagreb

Additional hard copies can be obtained from orders@intechopen.com

Dynamics of Arc Migration and Amalgamation - Architectural Examples from the NW Pacific Margin,
Authored by Yasuto Itoh, Osamu Takano, Reishi Takashima, Hiroshi Nishi and Takeyoshi Yoshida
p. cm.
Print ISBN 978-953-51-3221-9
Online ISBN 978-953-51-3222-6

For EU product safety concerns:
IN TECH d.o.o., Prolaz Marije Krucifikse Kozulić 3, 51000 Rijeka, Croatia,
info@intechopen.com or visit our website at intechopen.com.

Contents

Preface

This book provides readers with an overview of the dynamics of arc migration and amalgamation. The long-term activity of the Earth is governed by global plate motion. Subduction of oceanic plates not only causes ground tremors and magmatism but also controls material circulation around the surface of the globe. Our focus is on the NW Pacific margin where, strikingly, tectonic erosion and accretion phenomena simultaneously occur. It is an archetypal example of the evolutionary processes of convergent margins.

In Chapter 1, Takano presents a three-dimensional dissection of offshore basins around the Japanese Islands, utilizing extensive reflection on seismic data. His basin-filling stratigraphy helps identify unconformity events, stress regimes, succession trends, and development patterns for depositional systems, taking into consideration controlling factors such as plate tectonic conditions. It is a vision that leads to the concept of diverse and discontinuous evolution of sedimentary basins in convergent margins. Chapter 2 is devoted to paleomagnetic studies, which provide quantitative constraints on plate kinematics, but can often lead to cumbersome contradictions with the results of geological reconstructions. Itoh and Takashima attempt to determine reliable paleolatitudes and the northward migration history of crustal fragments in the NW Pacific by examining both previous datasets and newly obtained Mesozoic paleomagnetic directional data for Central Hokkaido. In addition to this geophysical analysis, in Chapter 3, Takashima provides stratigraphic and petrological insights into the early Cretaceous tectonic framework of the margin using the results of geochemical research. On the basis of a large amount of analytical data, he succeeds in the restoration of an ancient arc-trench system that had been mistakenly regarded as an accretionary complex. In Chapter 4, Itoh places new constraints on the late Cretaceous to Paleocene stress regimes in southwest Japan based on a microfabric analysis of a granitic pluton. It is an exquisite example of the application of rock magnetism to the visualization of fracture networks within rocks. Finally, in Chapter 5, we propose a plate reconstruction model for the NW Pacific region since 100 Ma and present a chronicle of tectonic episodes.

Thus, the contents of this book represent an interim report by authors who have engaged in a collaborative study of active margin tectonics for over a decade. At the same time, it sets a precedent for further exploration of mobile belt frontiers. Our hope is that the quantitative methodologies presented in this book, together with further efforts to integrate scientific knowledge, will pave the way for advancements in the comprehension of the diverse development processes associated with convergent margins.

Yasuto Itoh
Graduate School of Science, Osaka Prefecture University
Japan

Chapter 1

Intermittent Formation, Sedimentation and Deformation History of Cenozoic Forearc Basins along the Northwestern Pacific Margins as an Indicator of Tectonic Scenarios

Osamu Takano

Additional information is available at the end of the chapter

http://dx.doi.org/10.5772/intechopen.68290

Abstract

This chapter examines the basin-filling stratigraphy and major unconformity events of the Cenozoic forearc basins in the NE Japan, SW Japan, Ryukyu and Izu-Bonin forearc territories along the northwestern Pacific margins to obtain information on the background tectonic scenarios along the plate subduction zones. The forearc basin type and tectonic history are characteristic for each forearc territory, reflecting the differences in plate tectonic processes. Several major unconformity events seem to be synchronous for a forearc territory or whole forearc territories around Japan, suggesting that these events originated from more or less wider scale plate tectonic events. In the NE Japan forearc territory, the Oligocene unconformity can be the largest events, which transformed the forearc basin styles from the trench slope break-uplifted, fluvial system-dominated type to the tensional, deeper marine sloped type. In the SW Japan and Ryukyu forearc territories, the latest Oligocene to Middle Miocene gap was the transformation phase from the Palaeogene Shimanto-type forearc and accretionary complex, to the Neogene compressive, sloped to ridged forearc basins, developments of which have been interrupted by several unconformity events possibly related to changes in plate tectonic condition. These transformations of the forearc basin styles may reflect the changes in plate tectonic conditions in the northwestern Pacific region.

Keywords: forearc basin, unconformity event, basin-filling stratigraphy, northwestern Pacific margin, plate tectonics

1. Introduction

Forearc basin is a sedimentary basin developed between the volcanic arc and the plate subduction zone (**Figure 1**). Although the formation and subsidence mechanisms of forearc basins are still controversial [1, 2], the forearc basin behaviour including its formation, sedimentation, deformation and unconformable erosion may more or less reflect the plate subduction conditions along the plate convergent margins. In this scheme, it is supposed that the timing identification of such events provides important information on the plate tectonic history. Furthermore, the characterization of forearc basin-filling sediments by recognizing depositional-system transition trends, such as upward-shallowing and deepening successions associated with the balance between subsidence and sediment supply, may also provide crucial information on the basin setting parameters controlled by the plate tectonic conditions.

This chapter attempts to compile the timing of forearc basin formation, sedimentation and deformation along the northwestern Pacific margin, which has long been a plate convergent zone since the Mesozoic, to collect the information on the plate tectonic conditions during the Cenozoic. The examined forearc basin groups include the Sanriku-Hidaka-oki, Joban-oki, Boso, Tokai-oki-Kumano-nada, Shikoku-Miyazaki-oki, Ryukyu and Izu-Bonin forearc basins (**Figure 2**). For each forearc basin group, the lithostratigraphic division with sedimentation age determination, major depositional systems and successive trends of basin-filling sediments and major unconformities are detected with discussion of tectonic implications.

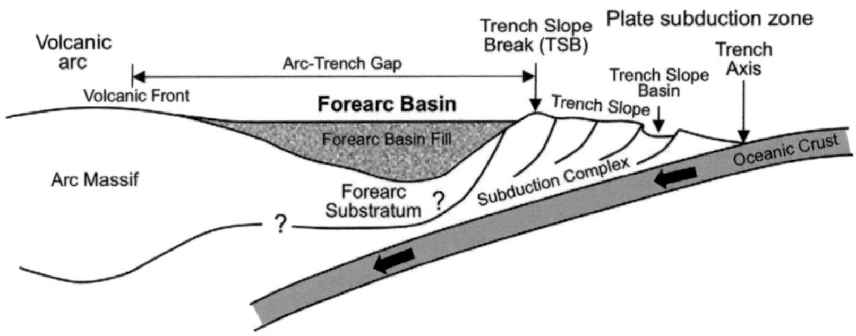

Figure 1. Schematic cross section of a forearc zone including a forearc basin, showing the basic terms used in this chapter. Modified after Refs. [2, 3].

2. Sanriku-Hidaka-oki forearc basins

2.1. Basin-filling stratigraphy

The Cenozoic Sanriku-Hidaka-oki forearc basins (basin group) are located along the N-S trending zone from the northeastern offshore of the Honshu Island to the south central part of Hokkaido (**Figures 2** and **3**). Basin-filling sedimentation in this forearc basin group started from the latest Jurassic, and has continued until the present with several intermission events.

Figure 4 demonstrates the Cenozoic stratigraphic correlation chart of the Hidaka-oki forearc basin on the northern part and the Sanriku-oki forearc basin on the southern part (**Figure 3**). In **Figure 4**, the Sanriku-oki forearc basin is further divided into the main part, which is mainly in the offshore territory at present, and the western basin margin part, where the Cretaceous to Palaeogene stratigraphic studies are well documented (e.g., [4–9]), using outcrop sections along the present coastline.

The Cenozoic sedimentation both in the Hidaka-oki and Sanriku-oki forearc basins began at around the Late Palaeocene time after the K/T gap unconformity (Tunc) event [10], and continued until the interruption by the large-scale Oligocene unconformity (Ounc [11]) formation event (**Figure 4**). Overall depositional systems in this stratigraphic interval between 'K-T gap Tunc' and 'Ounc' were coal-bearing fluvial systems with cyclic intercalations of bay-estuarine systems in response to relative sea-level changes (**Figure 4**: [12]). The detailed sedimentological basin analysis revealed that the depositional settings must have been in a restricted environment sheltered from an open marine condition, as shown in **Figure 5** [3, 13]. This forearc setting can be categorized as the shelved, benched type in the [2]'s forearc basin classification scheme, in which the trench slope break ridge is uplifted and the basin inside is overfilled with bay to fluvial sediments. Thus, the Cretaceous to Palaeogene forearc basins in

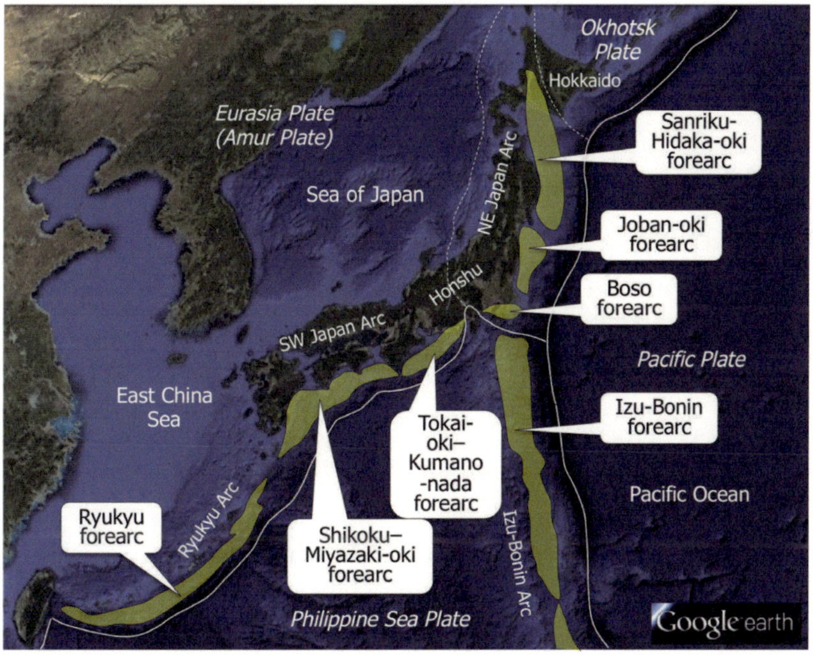

Figure 2. Map showing the distributions of forearc territories and groups in the western Pacific region. This study deals with four forearc territories: NE Japan, SW Japan, Ryukyu and Izu-Bonin forearc territories; and these forearc territories include several forearc basin groups as indicated in this figure. Image from Google Earth.

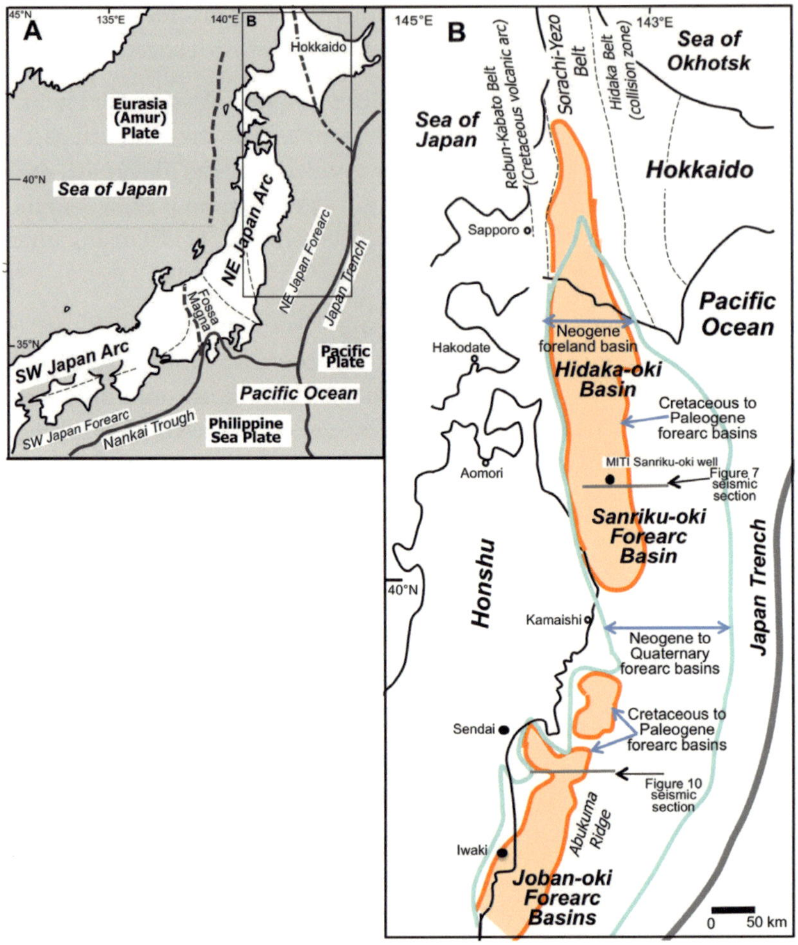

Figure 3. Location maps of the Sanriku-Hidaka-oki forearc basins and the Joban-oki forearc basins, showing the basin distributions and cross section locations. After [3, 11, 15]. Note that the basin distributions are different between the Cretaceous-Palaeogene forearc and the Neogene-Quaternary forearc, since the forearc basin type altered after the Oligocene unconformity (Ounc) event.

the Sanriku-Hidaka-oki forearc basins are characterized as such trench-slope-break uplifted, overfilled configuration with aggradational coal-bearing bay to fluvial sediments.

After the formation of 'Ounc', both the Hidaka-oki and Sanriku-oki forearc basins started drowning due to a wide-scale, tectonic hinge-type subsidence commencement (**Figure 6**), possibly related to the onset of tectonic erosion at the plate subduction zone along the NE Japan Arc. This event resulted in the transformation of the forearc basin type from the shelved, benched type to the sloped type of [2]'s scheme. The 'Ounc' unconformity surface was overlain by a transgressive succession, which was further followed by deeper marine muddy facies predominant over the Neogene and Quaternary successions in the Sanriku-oki forearc basin, although short-term episodes of unconformities (Munc: Miocene unconformity

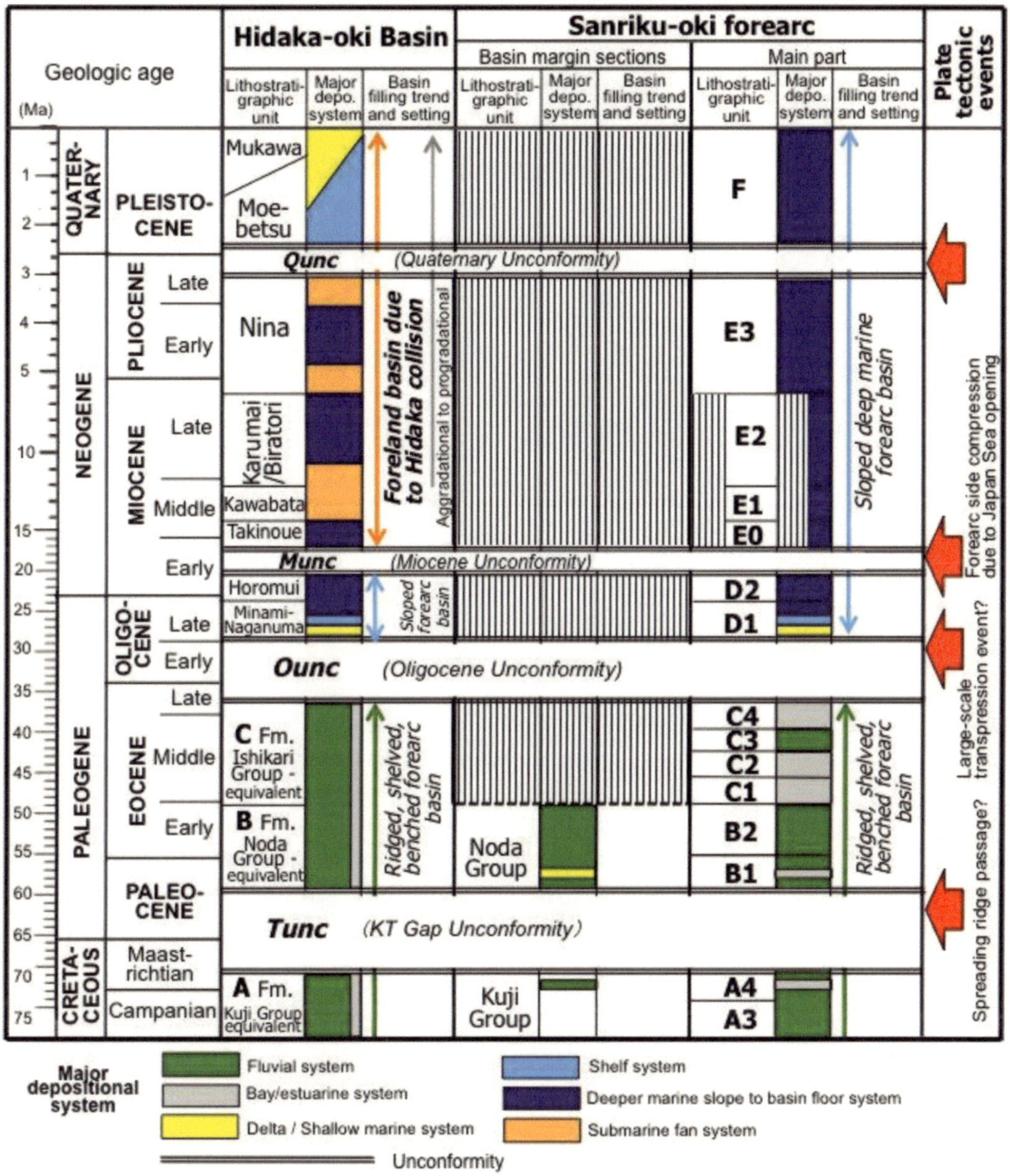

Figure 4. Cenozoic stratigraphic chart of the Sanriku-Hidaka-oki forearc basins, showing lithostratigraphy, major depositional systems, basin-filling trends and unconformity events as indicators of background plate tectonic history. Compiled after Refs. [5–7, 11, 14].

and Qunc: Quaternary unconformity) took place at around Early Miocene and the earliest Quaternary (**Figure 4**). On the contrary, the Hidaka-oki basin experienced the conversion from a forearc to foreland basin after the 'Munc' formation due to the westward collision of the Hidaka block associated with the Kuril forearc sliver migration (**Figure 6**). The deeper

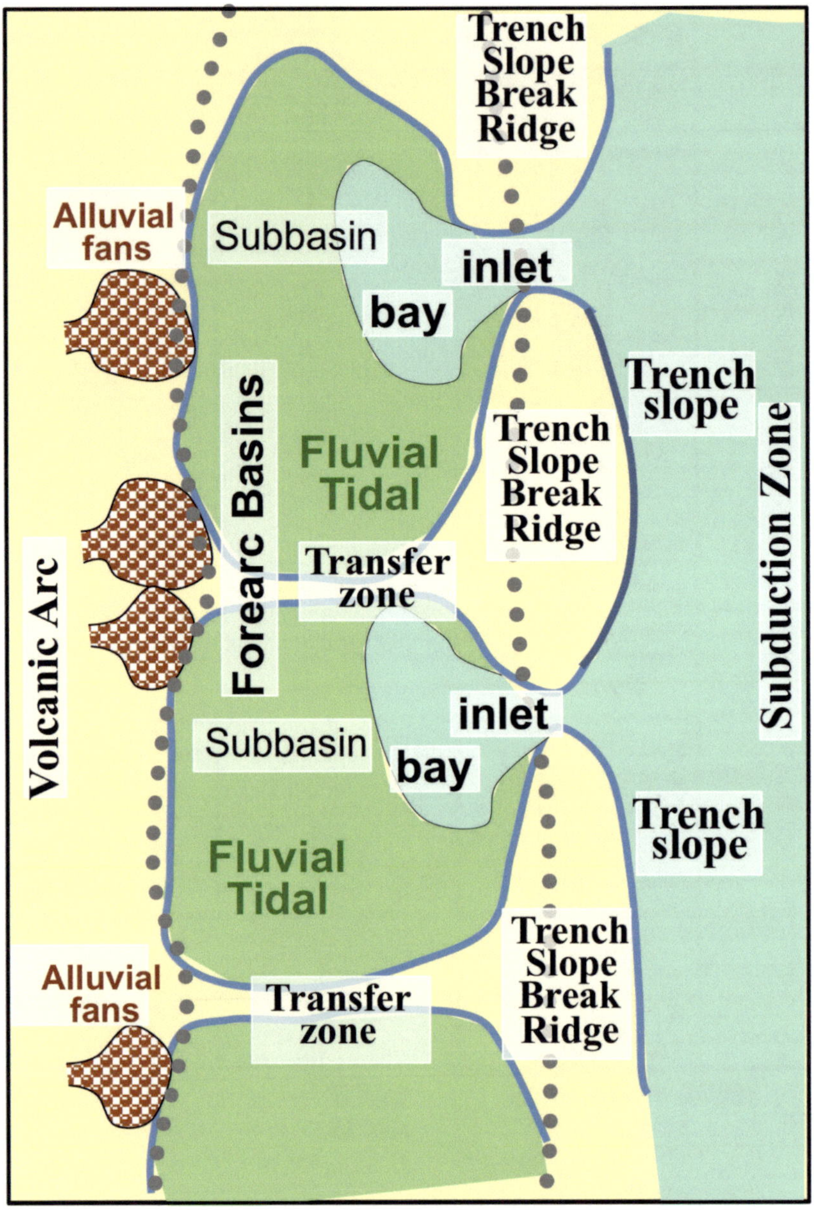

Figure 5. Schematic and conceptual forearc setting model for the Sanriku-Hidaka-oki forearc basins [3]. This setting can be categorized as the shelved, benched type of [2]'s forearc basin classification.

part of the foreland basin was filled with submarine-fan turbidites, which were derived from the uplifted Hidaka block (**Figure 4**). The basin-filling succession as a whole demonstrates an aggradational to upward-shallowing succession, resulting in the shelfal to shallow marine facies in the uppermost part (**Figure 4**).

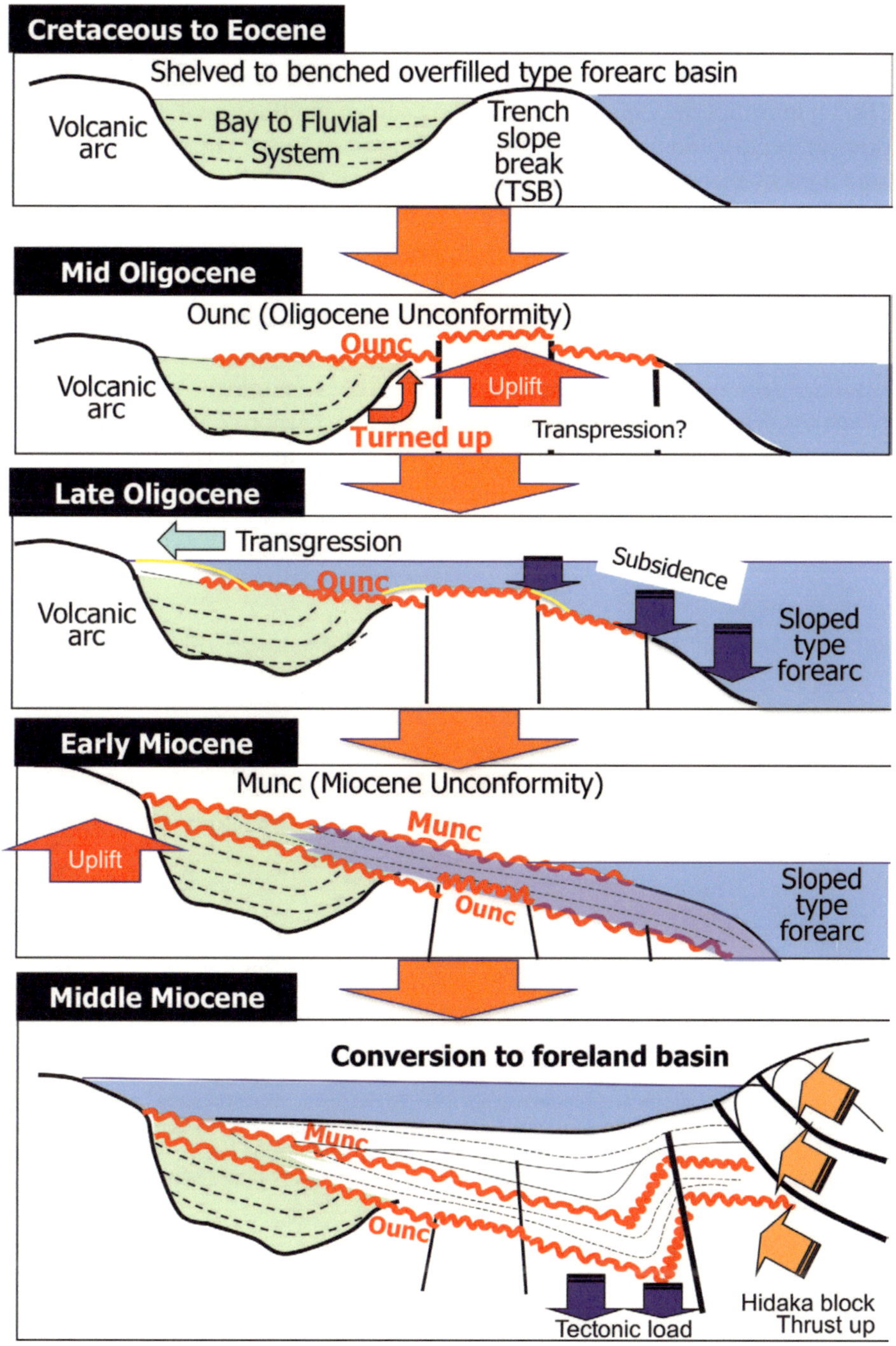

Figure 6. Cartoons showing the basin evolution and transformation of the Sanriku-Hidaka-oki forearc basins. The 'Ounc' event caused the transformation of the basin type from a benched, overfilled fluvial-dominated forearc to a sloped deeper marine forearc. The Hidaka-oki basin experienced further conversion to a foreland basin due to the Hidaka block collision.

2.2. Major unconformities and tectonic implications

In the Sanriku–Hidaka-oki forearc basins, four major unconformities: 'Tunc', 'Ounc', 'Munc' and 'Qunc' [11, 14] interrupt the Cenozoic basin-filling succession (**Figures 4** and **6**). 'Tunc (K-T gap unconformity)' occurs widely over the N-S trending zone including the Sanriku–Hidaka-oki forearc and Sorachi-Yezo belt in central Hokkaido (**Figure 3**) [5, 10], but the background tectonics is still controversial [5]. The largest Cenozoic tectonic event in the Sanriku-Hidaka-oki forearc basins is regarded as the formation of 'Ounc (Oligocene unconformity)', which corresponds to the turning point from the trench slope break-uplifted, shelved, benched, fluvial-dominated type forearc to the sloped, deeper marine forearc (**Figures 4** and **6**). The seismic survey section demonstrates a characteristic turned-up structure in the Cretaceous to Eocene forearc basin sediments due to intense arc-ward suppression and uplift of the trench slope break at the time of 'Ounc' formation (**Figures 6** and **7**). Since in the northern extension of this forearc zone in central Hokkaido, which has been defined as the 'Sorachi-Yezo belt', a strong dextral strike-slip tectonics was dominant during mid-Oligocene time [3, 13], it is considered that the 'Ounc' event more or less reflected the Oligocene dextral slip tectonics along the convergent zone.

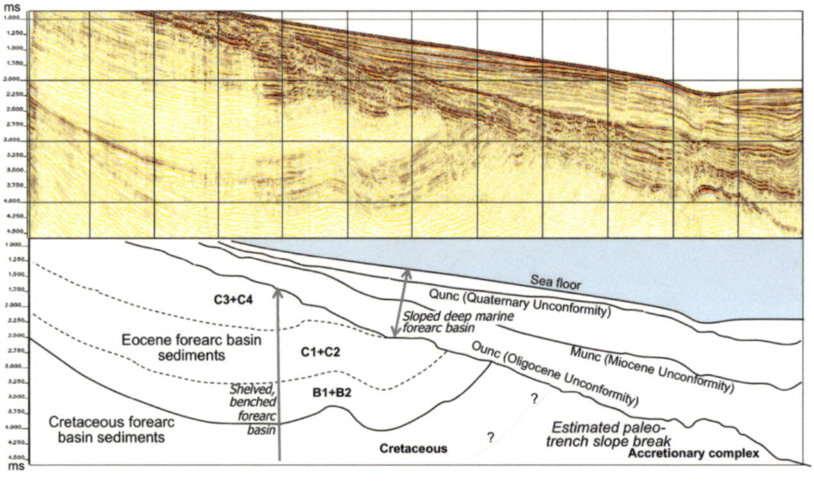

Figure 7. An E-W long 2D seismic section transecting the Sanriku-oki forearc basin, showing trench slope break uplift and Cretaceous to Eocene basin confinement (shelved, benched forearc) as modeled in **Figure 5**. After the formation of 'Ounc (Oligocene unconformity)', seaward dipping subsidence took place, and the forearc basin became a deep marine sloped type. The 2D seismic data were acquired in the MITI survey 'Shimokita-Kitakami' [16]. The seismic survey line location is shown in **Figure 3**. Modified after Ref. [3].

3. Joban-oki forearc basins

3.1. Basin-filling stratigraphy

The Joban-oki forearc basins (basin group) are located in the southern part of the forearc territory of the NE Japan arc (**Figure 3**). The outline of the basin-filling stratigraphy for the

http://dx.doi.org/10.5772/intechopen.68290

western onshore part has comprehensively been compiled by [17] with thorough consideration of the newest geologic age data set. The offshore stratigraphy has been reported by [15, 18, 19], using few well datasets including isotope and microfossil geologic age information. **Figure 8** demonstrates the basin-filling stratigraphy and major unconformity horizons both in the onshore and offshore parts of the Cenozoic Joban forearc basin. Although it seems that the Joban-oki basin-filling successions were interrupted more frequently by multiple unconformities, the total basin-filling pattern is similar to those of the Sanriku-Hidaka-oki forearc basins. The Cretaceous to Early Oligocene sediments below 'Ounc (Oligocene

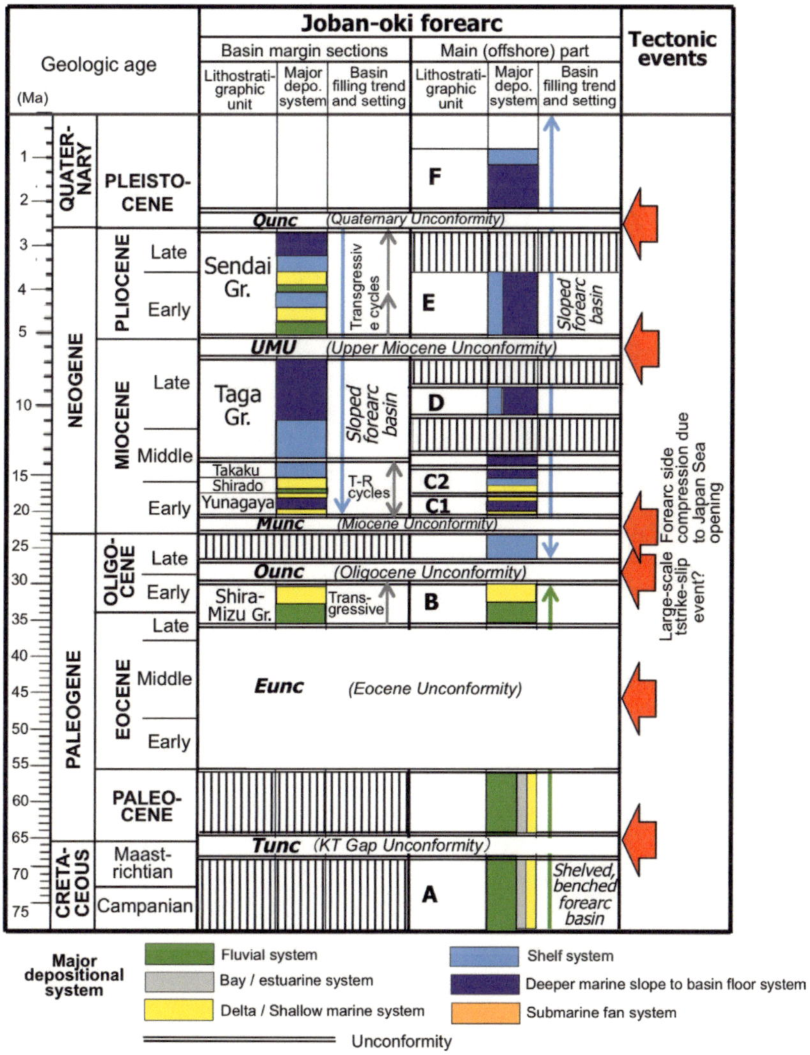

Figure 8. Cenozoic stratigraphic chart of the Joban-oki forearc basins, showing lithostratigraphy, major depositional systems, basin-filling trends and unconformity events as indicators of background plate tectonic history. Compiled after Refs. [15, 17–19].

unconformity)' are dominated by fluvial, delta and bay systems (**Figures 8** and **9**), and their distributions were limited to the arc-side of the Abukuma Ridge (**Figures 3**, **9** and **10**), suggesting that they were deposited in a restricted forearc basin inside the uplifted trench slope break. This situation is mostly the same as that of the Cretaceous to Eocene situation in the Sanriku–Hidaka-oki forearc basins (**Figure 5**), which were categorized as the shelved, benched type of [2]'s forearc basin classification.

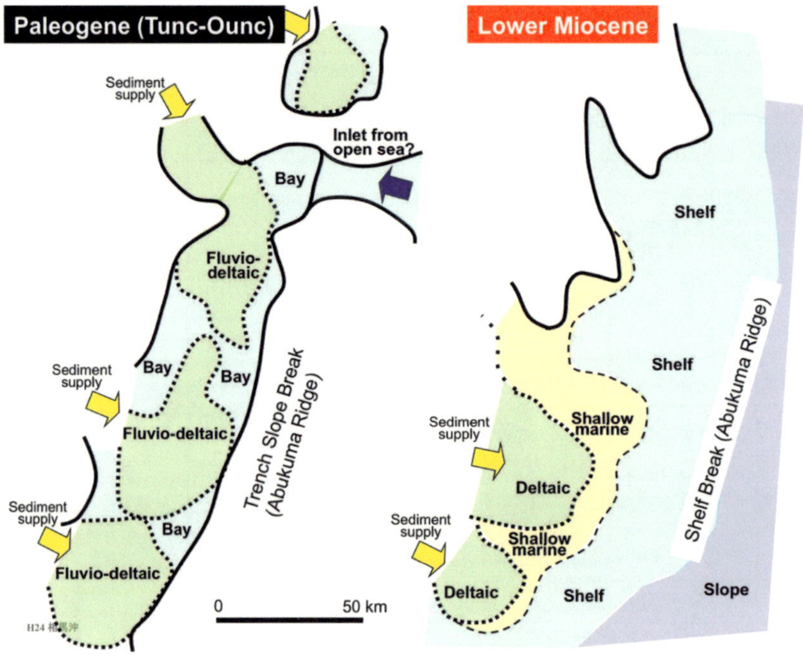

Figure 9. Schematic depositional models for the Paleogene and Lower Miocene in the Joban-oki forearc basins. Note that the Paleogene basin was restricted by a trench slope break ridge, whereas the Early Miocene basin faced directly to an open marine condition as a sloped forearc.

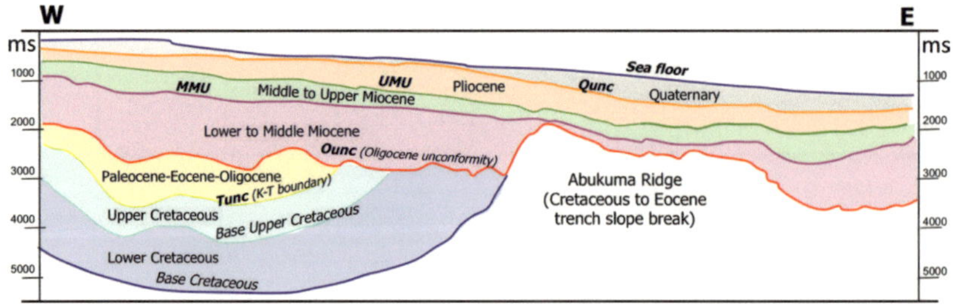

Figure 10. Interpreted transverse (E-W trending) seismic survey section in the northern part of the Joban-oki forearc basins. The used seismic survey section: MITI seismic survey 'Minami-Sanriku-Kashima-oki' M86-6 section [20]. The section line location is shown in **Figure 3**.

Based on the Palaeogene and Lower Miocene depositional models (**Figure 9**) and an E-W cross section of the Joban-oki basins (**Figure 10**), it is inferred that 'Ounc' functioned as the transitional event from a shelved, benched forearc to sloped forearc, as seen in the Sanriku-Hidaka-oki forearc basins. On the contrary to the Cretaceous to Eocene succession below 'Ounc', the Neogene and Quaternary succession above 'Ounc' shows an open marine condition, consisting mainly of a deeper marine slope system at the transgression phases and fluvial, delta, shallow marine to shelf systems at the regression phases. Since these Neogene to Quaternary sediments were deposited over the Cretaceous to Eocene trench slope break without any confinement in sediment distributions as shown in **Figures 9** and **10**, the Neogene to Quaternary Joban-oki forearc can be determined as the sloped type in [2]'s classification.

3.2. Major unconformities and tectonic implications

As the Sanriku-Hidaka-oki forearc basins, 'Tunc (K-T gap unconformity)', 'Ounc (Oligocene unconformity)', 'Munc (Miocene unconformity)' and 'Qunc (Quaternary unconformity)' can be recognized in the Joban-oki forearc basins. Among these unconformities, it seems that 'Ounc' was the largest event in the Joban-oki forearc history in terms of the transition from a trench slope break-uplifted type to a sloped type.

The Joban-oki forearc basins show additional unconformity events including the 'Eunc (Eocene unconformity)', 'Middle Miocene unconformity (MMU)' and 'Upper Miocene unconformity (UMU)' (**Figure 8**). Although the Eocene was one of the major depositional phases in the Sanriku-Joban-oki forearc basins, the Joban-oki forearc basins lack the major part of the Eocene because of this unconformity event. 'MMU' and 'UMU' also are characteristic in the Joban-oki forearc basins, which cannot be seen in the Sanriku-Hidaka-oki forearc basins.

4. Boso forearc basins

4.1. Basin-filling stratigraphy

The Boso forearc basins (basin group) are developed in the east side of the collision zone between the Izu-Bonin Arc and Honshu Arc (**Figure 11**). Since the major parts of the Boso forearc basins are exposed subaerially, many stratigraphic and sedimentologic outcrop studies have been conducted. Based on these studies, the basin-filling stratigraphy has comprehensively been outlined by [21] with the newest geologic age data set. **Figure 12** depicts the basin-filling stratigraphy and major unconformity horizons of the Boso forearc basins. The Early Miocene and older sediments in the Boso forearc basins have been preserved as accretionary prism-related sediments, which contain distorted, deformed blocks within muddy bedrocks. The Neogene forearc basins started at around the latest Early Miocene time, according to the newest geologic age data [21]. The Miura Group (or Awa Group) mainly consists of a transgressive succession at the base, and deeper marine slope to basin floor mudstones

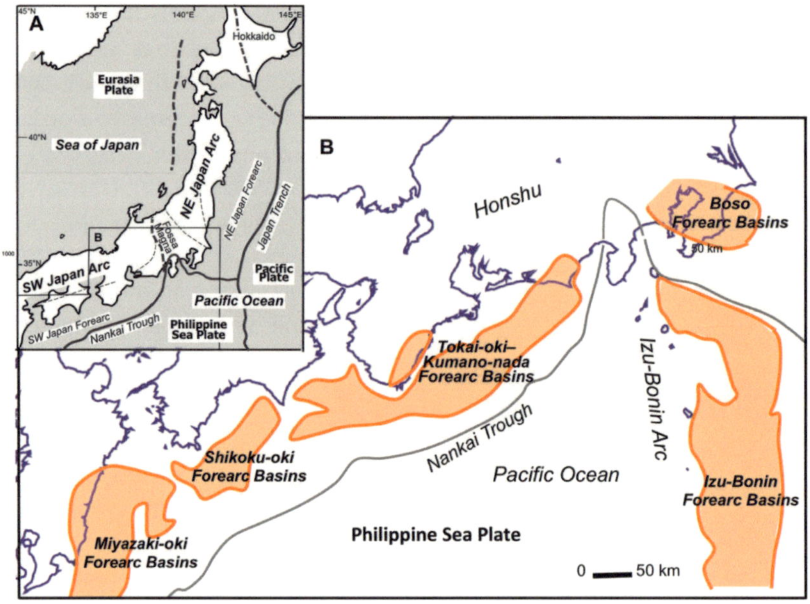

Figure 11. Location maps of the Boso, Tokai-oki-Kumano-nada, Shikoku-oki, Miyazaki-oki and Izu-Bonin forearc basins (forearc basin groups), showing the approximate basin distributions. After Refs. [23–25].

and turbidite sandstones in the main part. Ref. [22] reported that the submarine-fan turbidite sandstones of the middle to upper part of the Miura Group were deposited in a subduction zone-parallel, trough-like basin restricted by an uplifted trench slope break ridge, suggesting that at least Late Miocene trench slope break was uplifted and formed a ridged-type forearc basin configuration. After the formation of the 'Kurotaki unconformity', the latest Pliocene to Quaternary Kazusa forearc basin was created and the thick upward-shallowing succession of the Kazusa Group was deposited in the basin. The Late Pleistocene Shimosa Group unconformably overlies the Kazusa Group and was deposited in shallow marine and bay environments.

4.2. Major unconformities and tectonic implications

In the Cenozoic Boso forearc basins successions, the 'Early Miocene unconformity' between the accretionary prism complexes and overlying Miura Group, and the Late Pliocene 'Kurotaki unconformity' between the Miura and Kazusa Groups are the major unconformities in terms of plate tectonic history. The former 'Early Miocene unconformity' indicates the onset of the Neogene forearc basins with distinct formation of a trench slope break by thick piling of long-standing accretionary prism stacking. The latter Late Pliocene 'Kurotaki unconformity' can be suggestive of changes in plate subduction conditions, such as subduction direction change, which resulted in a large-scale uplift and subsequent subsidence of the forearc area.

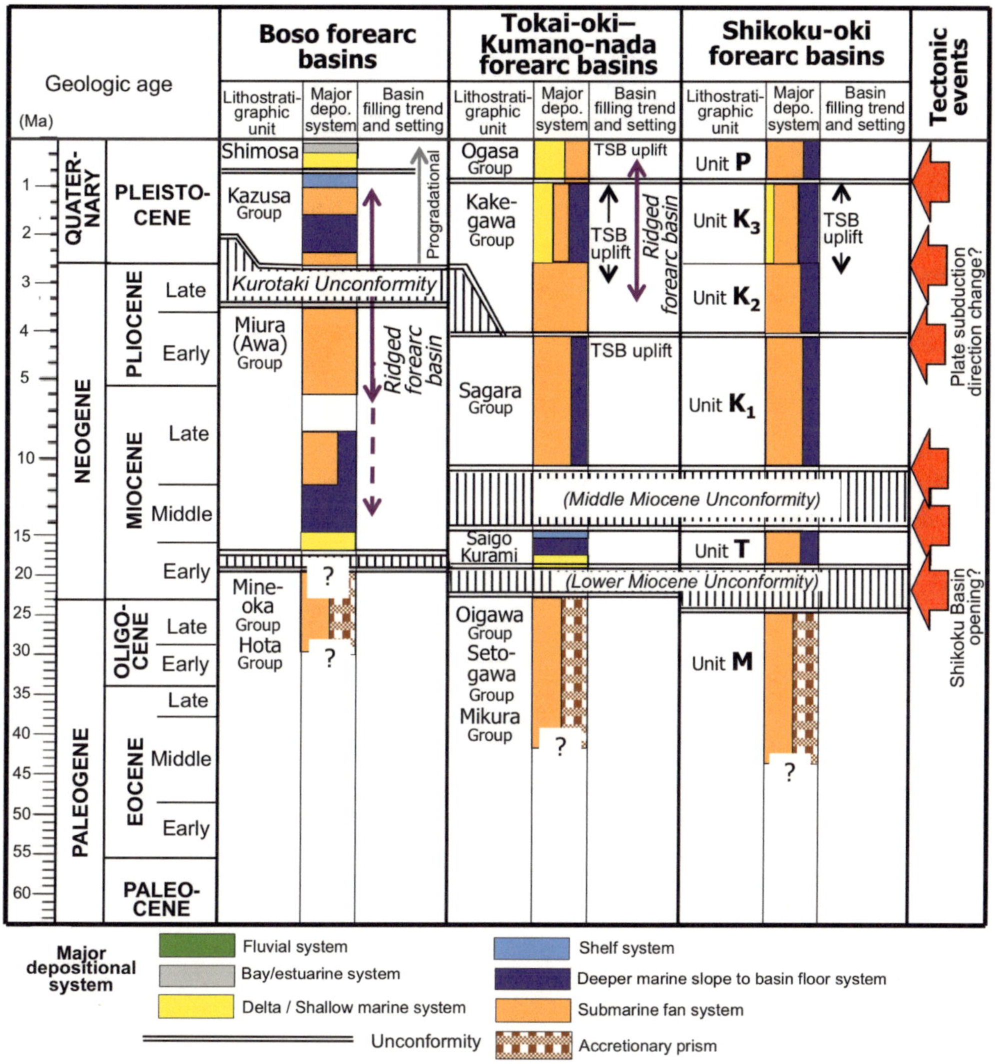

Figure 12. Cenozoic stratigraphic chart of the Boso, Tokai-oki-Kumano-nada and Shikoku-oki forearc basins, showing lithostratigraphy, major depositional systems, basin-filling trends and unconformity events as indicators of background plate tectonic history. Compiled after Refs. [3, 21, 24].

5. Tokai-oki-Kumano-nada forearc basins

5.1. Basin-filling stratigraphy

The Tokai-oki-Kumano-nada forearc basins (basin group) are developed in the eastern half of the SW Japan forearc territory along the Nankai Trough subduction zone, which is adjacent to the Honshu Arc-Izu-Bonin Arc collision zone at the eastern end (**Figure 11**). The basin-filling stratigraphy of the northeastern margin of the basins has been investigated using onshore outcrop sections [21, 26, 27], whereas the major part of the basin, which is occupied by the offshore territory, has been investigated using seismic survey and few exploration well dataset (e.g., [3, 28–32]).

Figure 12 depicts the basin-filling stratigraphy and major unconformity horizons of the Tokai-oki-Kumano-nada forearc basins. The Paleogene sediments, including the Oigawa, Setogawa and Mikura Groups, have been preserved as accretionary complex and related older forearc sediments, consisting mainly of submarine-fan turbidites and mudstones. The Neogene forearc basin sedimentation started during Early Miocene time, and the transgressive to regressive Saigo and Kurami Groups were deposited, first. The Neogene forearc basin sedimentation seems to have been interrupted several times by unconformity events possibly related to uplifting of trench slope break. The Middle Miocene unconformity separates the Saigo/Kurami Groups and the overlying Sagara Group, the Pliocene unconformity separates the Sagara Group and the overlying Kakegawa Group, and the Pleistocene unconformity separates the Kakegawa and the overlying Ogasa Group. **Figure 13** depicts the cyclic uplifting of trench slope break associated with forearc compression during the Pliocene to Pleistocene, suggesting such tectonic fluctuation, possibly related to change in plate subduction condition, created unconformity events in the Tokai-oki-Kumano-nada forearc basins.

5.2. Major unconformities and tectonic implications

As shown in **Figure 12**, it seems that the major unconformity events mostly match those in the Boso forearc basins, except the Middle Miocene unconformity, although unconformity timing shows a subtle difference.

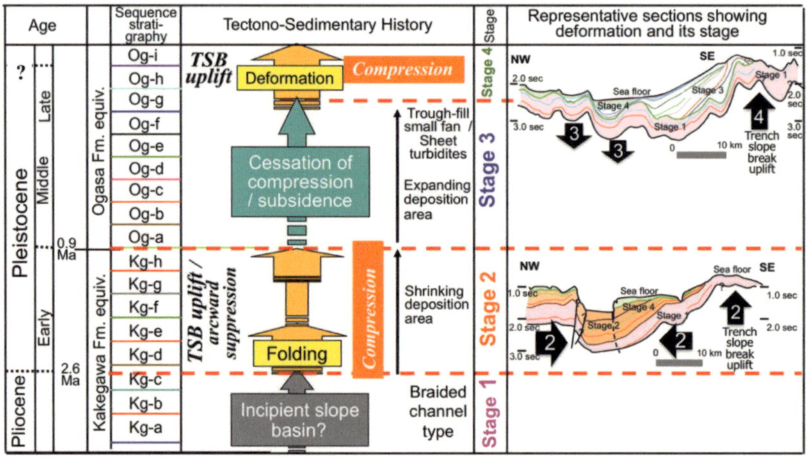

Figure 13. Stratigraphic tectono-sedimentary process chart of the Tokai-oki-Kumano-nada forearc basins, showing cyclic compression-relaxation fluctuation and associated basin deformation. Numbers in black arrows in the right column indicate deformation stages. Note that the climax phase of compression coincides in time with the unconformity horizon. Compiled and modified after Refs. [3, 31, 32].

6. Shikoku-oki forearc basins

6.1. Basin-filling stratigraphy

The Shikoku-oki forearc basins (basin group) are developed to the south of Shikoku Island along the Nankai Trough subduction zone (**Figure 11**). Since there are no outcrop sections and exploration wells penetrated, there is no direct information on the lithology in the Shikoku-oki basin sediments. However, seismic survey sections and their correlations to the adjacent areas provide the stratigraphic outline of the Shikoku-oki basin sediments. Ref. [24] described the outline of the basin-filling stratigraphy and sediment distributions using offshore seismic survey data set. According to [24], the basin-filling sediments can be divided into Unit M, T, K_1, K_2, K_3 and P in a stratigraphic ascending order **Figure 12**. The acoustic basement corresponds to Palaeogene Unit M, and is considered to have been composed of accretionary complex. Miocene Unit T is widely distributed, and is considered to be equivalent to the Middle Miocene Tanabe, Saigo and Kurami Groups. Upper Miocene to Lower Pliocene Unit K_1 overlies depressions formed at the Middle Miocene unconformity event, and is considered to be equivalent to the Sagara Group. Pliocene-Pleistocene Units K2 and K3 seem to be affected by compressive undulation events, and is considered to be equivalent to the Kakegawa Group. Pleistocene Unit P blankets the underlying sediments, and is considered to be equivalent to the Ogasa Group.

6.2. Major unconformities and tectonic implications

As shown in **Figure 12**, it seems that the major unconformity events mostly coincide in time with those in the Tokai-oki-Kumano-nada forearc basins: Lower Miocene unconformity between the accretionary complex and Neogene forearc basin initiation sediments, Middle Miocene unconformity, Pliocene unconformity and Pleistocene unconformity. All these are regarded as common tectonic events along the Nankai Trough during the Neogene time.

7. Miyazaki-oki forearc basins

7.1. Basin-filling stratigraphy

The Miyazaki-oki forearc basins (basin group) are located on the forearc side of the junction between the Honshu (SW Japan) Arc and Ryukyu Arc (**Figure 11**). The basin-filling stratigraphy of the northwestern margin of the basins has been investigated using onshore outcrop sections by many researchers (e.g., [21, 33-35]) with detailed geologic age controls, whereas the major part of the basin, which is occupied by offshore territory, has been investigated using seismic survey and few exploration well data set [23].

Figure 14 depicts the basin-filling stratigraphy and major unconformity horizons of the Miyazaki-oki forearc basins. The Cenozoic basin-filling successions of the Miyazaki-oki forearc basins seem to be separated by the Early Miocene to early Late Miocene unconformity or hiatus into the lower accretionary prism-related older sediments and the upper Neogene forearc basin fill. The lower accretionary prism-related sediments, including the Paleogene Nichinan, Hyuga, Gumage and Kitakawa Groups, are composed of contorted and blocked accretionary complexes and the adjacent forearc basin sediments consisting mainly of gravity flow deposits and deeper marine shales. The upper Neogene forearc basin fill includes the Upper Miocene to Pliocene Miyazaki Group and Pleistocene Hyuganada Group, and mainly consists of large-scale transgressive-regressive cycles of fluvial, delta, shallow marine, shelf, delta-fed submarine-fan and slope sediments.

7.2. Major unconformities and tectonic implications

As shown in **Figure 14**, the Early Miocene to early Late Miocene gap is the largest unconformity or hiatus separating the Miyazaki-oki forearc basin-filling successions. Smaller unconformities can be observed within the Neogene forearc basin fill; those are, the base of the Miyazaki and Tsuma facies (latest Miocene) and the boundary between the Miyazaki and Hyuga-nada Groups (earliest Pleistocene).

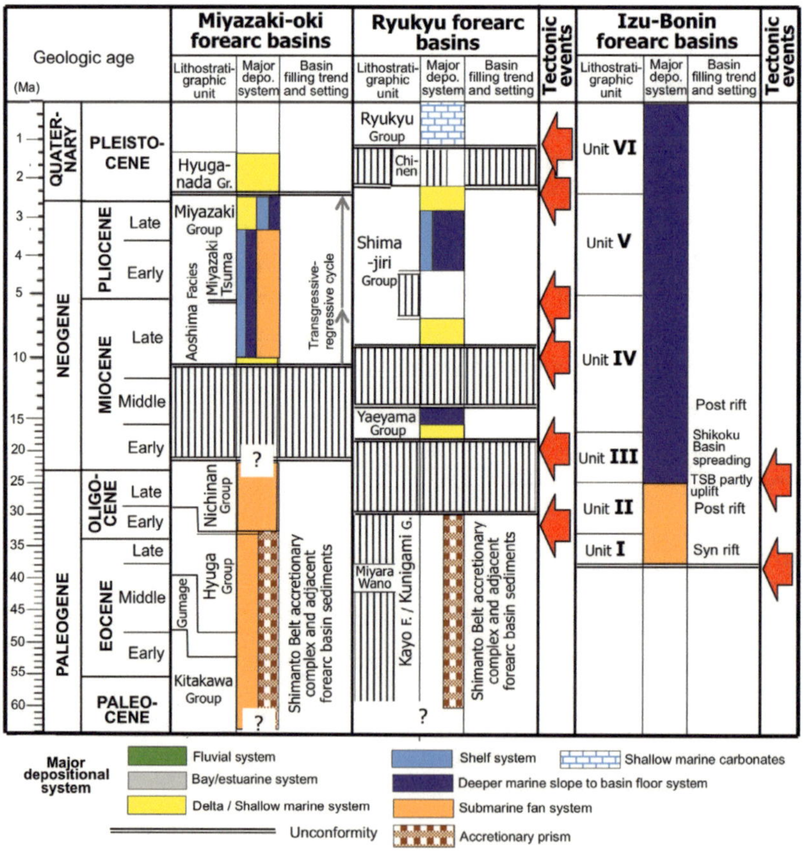

Figure 14. Cenozoic stratigraphic chart of the Miyazaki-oki, Ryukyu, Izu-Bonin forearc basins, showing lithostratigraphy, major depositional systems, basin-filling trends and unconformity events as indicators of background plate tectonic history. Compiled after Refs. [21, 23, 25, 33–35].

8. Ryukyu forearc basins

8.1. Basin-filling stratigraphy

The Ryukyu forearc basins (basin group) are located on the forearc side of the Ryukyu Arc (**Figure 2**). The basin-filling stratigraphy has been investigated using partly exposed onshore outcrop sections in the archipelagos, offshore seismic survey results and several exploration well dataset (e.g. [23, 35]).

Figure 14 depicts the basin-filling stratigraphy and major unconformity horizons of the Ryukyu forearc basins. It seems that the Cenozoic Ryukyu forearc basin developments have been sporadic in time and space, being separated by several unconformities/hiatuses and forearc territory segments. The Palaeogene accretionary complex and associated forearc sediments had been deposited until mid-Oligocene time. After the no deposition period during the Late Oligocene to Early Miocene, sporadic sedimentation of the Yaeyama Group and its equivalent took place in the Early to Middle Miocene. Thick accumulation of the Shimajiri Group started at around mid-Late Miocene, showing a transgression-regression cycle with short hiatus periods in the middle. In association with the Okinawa Trough opening, which prevented siliciclastic input, the Upper Quaternary Ryukyu Group, mainly consisting of shallow marine carbonate rocks, was deposited upon the early Quaternary unconformity.

8.2. Major unconformities and tectonic implications

As shown in **Figure 14**, the 'Late Oligocene-Early Miocene gap' is the largest unconformity or hiatus separating between the underlying accretionary complex and the Neogene forearc basin fill. The Middle to Late Miocene unconformity separates the sporadic developments of the Middle Miocene Yaeyama Group and the overlying Upper Miocene to Pliocene Shimajiri Group. The Early Pleistocene unconformity between the Shimajiri and Ryukyu Groups is considered to be related to the Okinawa Trough opening.

9. Izu-Bonin forearc basins

The Izu-Bonin forearc basins (basin group) are located along the forearc side of the Izu-Bonin Arc (**Figures 2** and **11**). Although the detailed basin-filling stratigraphy has not been investigated, fine meshes of 2D seismic survey sections and some ODP drilling data have revealed the stratigraphic framework and major lithology information in the Izu-Bonin forearc basins (e.g. [25, 36]).

Figure 14 depicts the outlined basin-filling stratigraphy of the Izu-Bonin forearc basins. According to [25], the forearc basin fill can be divided into Units I–VI in a stratigraphic ascending order. As the initiation of the forearc basins, Unit I was deposited at the bottom of the forearc basins as syn-rift deposits during the Eocene time. Unit II, which was deposited during the Early to mid-Oligocene, conformably overlies Unit I, and is considered as post-rift deposits. Both Units I and II mainly consist of turbidite sandstones. Unit III is regarded as the coincident sediments with the Shikoku Basin opening event, and partly shows unconformable relationship with the underlying Unit II. Although the depocenter shifts can be observed, there are no distinct event features in the succession of Units IV–VI, which are mainly composed of hemipelagic shales and volcaniclastics derived from the adjacent volcanic arc.

10. Synthesis: possible event phase tracing

As a result of compilation above, **Figure 15** summarizes the unconformity events and basin types of the Sanriku-Hidaka-oki, Joban-oki, Boso, Tokai-oki-Kumano-nada, Shikoku-oki, Miyazaki-oki, Ryukyu and Izu-Bonin forearc basins. Based on **Figure 15**, some characteristic features can be suggested as follows.

10.1. Difference between the NE Japan, SW Japan, Ryukyu and Izu-Bonin forearc basins

Figure 15 indicates that the forearc basin type and tectonic history are different between the NE Japan, SW Japan-Ryukyu and Izu-Bonin forearc territories.

The Cretaceous to Palaeogene forearc basin types in the NE Japan forearc territory are characterized by the shelved, benched, fluvial-dominated type, which was followed by the deeper marine sloped type after the 'Ounc' formation, except for the Hidaka-oki basin, which was transferred to a foreland basin. This deeper marine sloped type forearc basin shows overall subsidence of the forearc regions, including the underlying benched-type forearc basins, into

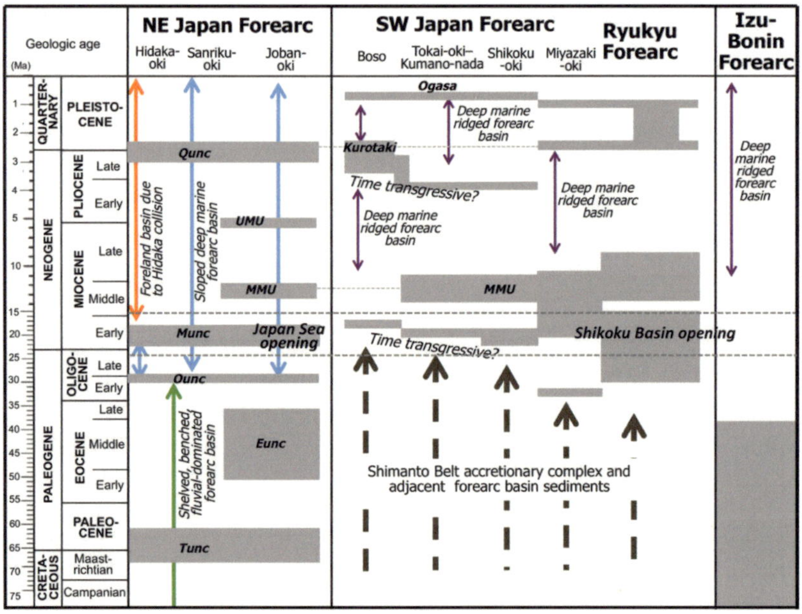

Figure 15. Summary chart of unconformity event phases and basin type transitions in the Sanriku-Hidaka-oki, Joban-oki, Boso, Tokai-oki-Kumano-nada, Shikoku-oki, Miyazaki-oki, Ryukyu and Izu-Bonin forearc basins. Gray territories show sedimentation gaps formed by unconformity or hiatus indicative of tectonic events.

a deeper marine condition, and can be classified as the tensional non-accretion forearc type of [37]. On the contrary, the Neogene forearc basin types in the SW Japan-Ryukyu and Izu-Bonin forearc territories are characterized mainly by the deeper marine sloped or ridged type, depending on the arc-ward suppression of accretionary complex related to active accretion in the plate subduction zone. This type of forearc basins can be classified as the compressive accretion type of [37].

With regard to the tectonic history, **Figure 15** suggests that the NE Japan, SW Japan-Ryukyu and Izu-Bonin forearc territories experienced different tectonic history. The largest tectonic event in the NE Japan forearc can be 'Ounc (Oligocene unconformity)', which caused the transformation of the basin types from the shelved, benched, fluvial-dominated type to the deeper marine sloped type. On the contrary, in the SW Japan-Ryukyu and Izu-Bonin forearc territory, the Shimanto Belt-related accretionary complex formation and related forearc basin sedimentation continued until the Late Oligocene to Early Miocene, and subsequently, the Neogene forearc basin sedimentation started at the Early, Middle or Late Miocene. The Izu-Bonin forearc basin started at the Eocene and has continued sedimentation until the present.

10.2. Unconformity synchronicity indicating plate tectonic events

Figure 15 also indicates that some unconformity events are synchronous for a forearc territory or whole forearc territories around Japan, suggesting that these events originated from more or less wider-scale plate tectonic events. 'Ounc (Oligocene unconformity)' can clearly be observed along the NE Japan forearc territory, and is considered to be caused by a large-scale plate tectonic event as indicated by the erosional turned-up structure and transformation of the forearc basin style. 'Munc (Miocene unconformity)' in the NE Japan forearc territory and the 'Early Miocene unconformities' developed in the SW Japan forearc territory can be involved within the opening event periods of the Shikoku Basin and the Sea of Japan, suggesting that these unconformities reflected the backarc opening tectonics. The 'Middle Miocene unconformities (MMU)' also demonstrate synchronous formation both in the NE Japan and SW Japan forearc territories. 'Qunc (Quaternary unconformity)' in the NE Japan forearc territory possibly coincided in time with the Kurotaki unconformity event in the SW Japan forearc territory.

10.3. Unconformity event diachronicity

Although synchronicity of unconformity events can be suggested in a wider view, some unconformity events may indicate systematic diachronicity in a finer view.

'Early Miocene unconformities' in the SW Japan forearc-Ryukyu forearc territories show a time transgressive occurrence of each unconformity event from the west to the east, indicating the unconformity event took place slightly earlier in the west and later in the east. Since this 'Early Miocene unconformity' phases coincide in time with the Shikoku Basin opening, this diachronicity is possibly related to the Shikoku Basin opening event.

The late Pliocene unconformities also show a time transgressive pattern, indicating unconformity occurred earlier in the west and later in the east.

11. Conclusions

This chapter attempted to compile the basin-filling stratigraphy and major unconformity events of the Cenozoic forearc basins in the Sanriku-Hidaka-oki, Joban-oki, Boso, Tokai-oki-Kumano-nada, Shikoku-oki, Miyazaki-oki, Ryukyu-oki and Izu-Bonin forearc territories. The results of the compilation provided the following new findings.

(1) The forearc basin type and tectonic history are characteristic of the NE Japan, SW Japan-Ryukyu and Izu-Bonin forearc territories, respectively, reflecting the differences in plate tectonic processes.

(2) Several major unconformity events were synchronous for a forearc territory or whole forearc territories around Japan, suggesting that these events originated from more or less wider-scale plate tectonic events.

(3) 'Ounc (Oligocene unconformity)' is considered as the largest plate tectonic event in the NE Japan forearc territory, which converted the forearc basin type from the benched fluvial-dominated type to the deeper marine sloped type.

(4) In the SW Japan and Ryukyu forearc territories, the latest Oligocene to Middle Miocene gap was the transformation phase from the Paleogene Shimanto-type forearc and accretionary complex to the Neogene compressive, sloped to ridged forearc basins.

(5) 'Munc (Miocene unconformity)' in the NE Japan forearc territory and the 'Early Miocene unconformities' developed in the SW Japan forearc territory can be related to the opening event of the Shikoku Basin and the Sea of Japan.

(6) Some unconformity events may indicate systematic diachronicity in a finer view, possibly reflecting gradual shifts of a tectonic event.

Acknowledgments

The author is grateful for JAPEX, JOGMEC and METI, who gave data permission for publication.

Author details

Osamu Takano

Address all correspondence to: osamu.takano@japex.co.jp

Technical Division, Japan Petroleum Exploration (JAPEX) Research Center, Tokyo, Japan

References

[1] Dickinson WR, Seely DR. Structure and stratigraphy of forearc regions. American Association of Petroleum Geologists Bulletin. 1979;**63**:2–31

[2] Dickinson WR. Forearc basins. In: Busby C, Ingersoll RV, editors. Tectonics of Sedimentary Basins. Oxford: Blackwell; 1995. pp. 221–261

[3] Takano O, Itoh Y, Kusumoto S. Variation in forearc basin configuration and basin-filling depositional systems as a function of trench slope break development and strike-slip movement: examples from the Cenozoic Ishikari-Sanriku-oki and Tokai-oki-Kumano-nada forearc basins, Japan. In: Itoh Y, editor. Mechanism of Sedimentary Basin Formation: Multidisciplinary Approach on Active Plate Margins. ISBN: 978-953-51-1193-1, InTech; Rijeka, Croatia. 2013. pp. 3–25. Available from: http://www.intechopen.com/books/mechanism-of-sedimentary-basin-formation-multidisciplinary-approach-on-active-plate-margins/variation-in-forearc-basin-configuration-and-basin-filling-depositional-systems-as-a-function-of-

[4] Terui K, Nagahama H. Depositional facies and sequences of the Upper Cretaceous Kuji Group, Northeast Japan. Memoir of the Geological Society of Japan. 1995;**45**:238–249 (in Japanese with English abstract)

[5] Ando H. Stratigraphic correlation of Upper Cretaceous to Paleocene forearc basin sediments in Northeast Japan: cyclic sedimentation and basin evolution. Journal of Asian Earth Sciences. 2003;**21**:919–933

[6] Ando H. Upper Cretaceous fore arc basin sequences in Northeast Japan: Large-scale controlling factors as eustasy, volcanism and relative plate motion. Journal of the Geological Society of Thailand. 2004;1:35–44

[7] Ando H. Geologic setting and stratigraphic correlation of the Cretaceous to Paleocene Yezo forearc basin in Northeast Japan. Journal of Japanese Association for Petroleum Technology. 2005;**70**:24–36 (in Japanese with English abstract)

[8] Ando H. Apparent stacking patterns of depositional sequences in the Upper Cretaceous shallow marine to fluvial successions, Northeast Japan. Memoir of the Geological Society of Japan. 1997;**48**:43–59

[9] Umetsu K, Hirayama R, Sonoda T, Takashima R. Marine and nonmarine deposits of the Cretaceous Miyako and Kuji groups and Late Cretaceous terrestrial vertebrates in Iwate Prefecture, northeast Japan. Journal of the Geological Society of Japan. 2013;**119**(Supplement);82–95 (in Japanese with English abstract)

[10] Ando H, Tomosugi T. Unconformity between the Upper Maastrichtian and Upper Paleocene in the Hakobuchi Formation, north Hokkaido, Japan: A major time gap within the Yezo forearc basin sediments. Cretaceous Research. 2005;**26**:85–95

[11] Osawa M, Nakanishi S, Tanahashi M, Oda H. Structure, tectonic evolution and gas exploration potential of offshore Sanriku and Hidaka provinces, Pacific Ocean, off northern Honshu and Hokkaido, Japan. Journal of Japanese Association for Petroleum Technology. 2002;**67**:38–49 (in Japanese with English abstract)

[12] Takano O, Waseda A. Sequence stratigraphic architecture of a differentially subsiding bay to fluvial basin: The Eocene Ishikari Group in the Ishikari Coal Field, central Hokkaido, Japan. Sedimentary Geology. 2003;**160**(1–3):131–158

[13] Itoh Y, Takano O, Kusumoto S, Tamaki M. Mechanism of longstanding Cenozoic basin formation in central Hokkaido: An integrated basin study on an oblique convergent margin. Progress in Earth Planetary Sciences. 2014;**1**:6.

[14] Japan National Oil Corporation (JNOC). Report for the geological study of the MITI Sanriku-oki well. Fiscal Year 1998. Tokyo: Japan National Oil Corporation; 2000. p. 48

[15] Iwata T, Hirai A, Inaba T, Hirano M. Petroleum system in the Offshore Joban Basin, northeast Japan. Journal of the Japanese Association for Petroleum Technology. 2002;**67**:62–71 (in Japanese with English abstract)

[16] Japan National Oil Corporation. Report on the geophysical survey 'Shimokita-Kitakami'. Fiscal Year 1973. Tokyo: Japan National Oil Corporation; 1974.

[17] Ando H, Yanagisawa Y, Komatsubara J. Cretaceous to Neogene strata in the Joban area and their sedimentation of the Joban forearc basin. Journal of Geological Society of Japan. 2011;**117**(Supplement):49–67

[18] Kato S, Akiba F, Moriya S. The Upper Cretaceous-Cenozoic stratigraphy and geologic structure in the offshore Soma area, northeast Japan. Journal of the Geological Society of Japan. 1996;**102**:1039–1051

[19] Kato S, Tazawa K, Nakano T. Cenozoic strontium isotope stratigraphy in the MITI Soma-oki well, northeastern Japan. Journal of the Geological Society of Japan. 1997; **103**:1046–1052

[20] Japan National Oil Corporation. Report on the geophysical survey 'Minami-sanriku–Kashima-oki'. Fiscal Year 1986. Tokyo: Japan National Oil Corporation; 1987

[21] Ito M, Kameo K, Satoguchi Y, Masuda F, Hiroki Y, Takano O, Nakajima T, Suzuki N. Neogene–Quaternary sedimentary successions. In: Moreno T, Wallis S, Kojima T, Gibbons W, editors. The Geology of Japan. London: Geological Society; Special Publication. 2016. pp. 313–341

[22] Tokuhashi S. Sedimentological study of the flysch type alteration of Hk horizon in the Kiyosumi Formation (Part 2). Depositional processes and circumstances of sandstone beds. Journal of the Geological Society of Japan. 1976;**82**:757–764

[23] Association of Natural Gas Mining and Association for offshore petroleum exploration, editor. Petroleum and Natural Gas Resources of Japan. Tokyo: Association of Natural Gas Mining and Association for offshore petroleum exploration; 1992. p. 520 (in Japanese)

[24] Sasaki K, Okuda Y, Kato S. Petroleum geology of the Pacific side of Southwest Japan. In: Association of Natural Gas Mining and Association for Offshore Petroleum Exploration, editor. Petroleum and Natural Gas Resources of Japan. Tokyo: Association of Natural Gas Mining and Association for Offshore Petroleum Exploration; 1992. pp. 225–246

[25] Hoshi K, Yanagimoto Y, Akiba F, Koda K. Architecture and tectonic evolution of sedimentary basins in the Izu-Ogasawara Arc based on a seismic reflection survey interpretation. Journal of Geography (Chigaku Zasshi). 2015;**124**:847–876 (in Japanese with English abstract)

[26] Tsukawaki S. Depositional environments of the Sagara and Kakegawa Groups (Middle Miocene-Early Pleistocene), and the evolution of the sedimentary basin, central Japan. Science Report of the Tohoku University, Sendai, Second Series (Geology). 1994;**63**:1–38

[27] Kameo K. Upper Neogene and Quaternary stratigraphy in the Kakegawa district based on the calcareous nannofossil datum planes, with reference to the stratigraphic position of the Tamari Formation. Journal of the Geological Society of Japan. 1998;**104**:672–686

[28] Japan National Oil Corporation. Report for the MITI offshore geophysical survey Tokai-oki–Kumano-nada. Fiscal Year 2001. Tokyo: Japan National Oil Corporation; 2003

[29] Japan National Oil Corporation. Report for the geological study of the MITI Tokai-oki–Kumano-nada wells. Fiscal Year 2003. Tokyo: Japan National Oil Corporation; 2004

[30] Arai K, Okamura Y, Ikehara K, Ashi J, Soh W, Kinoshita M. Active faults and tectonics on the upper forearc slope off Hamamatsu City, central Japan. Journal of the Geological Society of Japan. 2006:**112**;749–759 (in Japanese with English abstract)

[31] Takano O, Nishimura M, Fujii T, Saeki T. Sequence stratigraphic distribution analysis of methane-hydrate-bearing submarine-fan turbidite sandstones in the Eastern Nankai Trough area: relationship between turbidite facies distributions and BSR occurrence. Journal of Geography. 2009;118:776–792 (in Japanese with English abstract)

[32] Takano O, Fujii T, Saeki T, Shimoda N, Noguchi S, Nishimura M, Takayama T, Tsuji T. Application of sedimentological methodology to the methane-hydrate exploration project in the eastern Nankai Trough area. Journal of Japanese Association of Petroleum Technology. 2010;**75**:9–19 (in Japanese with English abstract)

[33] Nakamura Y, Ozawa T, Nobuhara T. Stratigraphy and molluscan fauna of the upper Miocene to lower Pliocene Miyazaki Group in the Aoshima area, Miyazaki Prefecture, southwest Japan. Journal of the Geological Society of Japan. 1999;**105**:45–60 (in Japanese with English abstract)

[34] Oda M, Chiyonobu S, et al. Integrated magnetostratigraphy of the Pliocene-Pleistocene Miyazaki succession, southern Kyushu, southwest Japan: Implications for an early Pleistocene hiatus and defining the base of the Gelasian (P/P boundary section) in Japan. Journal of Asian Earth Sciences. 2001;**40**:84–97

[35] Miyazaki K, Ozaki M, Saito M, Toshimitsu S. The Kyushu-Ryukyu Arc. In: Moreno T, Wallis S, Kojima T, Gibsons W, editors. The Geology of Japan. London: Geological Society; 2016. pp. 139–174

[36] Taylor B. Rifting and the volcanic-tectonic evolution of the Izu-Bonin-Mariana Arc. Proceedings of ODP, Scientific Results. 1992;**126**:627–651

[37] Noda A. Forearc basins: Types, geometries, and relationships to subduction zone dynamics. Geological Society of America Bulletin. 2016;**128**:879–895

Chapter 2

Cretaceous Research: Paleolatitudes and Northward Migration of Crustal Fragments in the NW Pacific Inferred from Paleomagnetic Studies

Yasuto Itoh and Reishi Takashima

Additional information is available at the end of the chapter

http://dx.doi.org/10.5772/67359

Abstract

Lateral migration of the Oshima and Sorachi-Yezo Belts within south central Hokkaido was quantitatively evaluated by means of paleomagnetic analyses in order to identify allochthonous blocks on the northwestern Pacific margin. The remanence stability of the Late Jurassic to Early Cretaceous voluminous igneous succession of the Kumaneshiri and Sorachi Groups and the overlying forearc sediments of the Cretaceous Yezo Group was evaluated through rock magnetic experiments. Twelve of the sites yielded characteristic primary components residing in mixtures of titanomagnetite and hematite having various mixing ratios. After an appropriate correction of inclinations' shallowing of the post-depositional detrital remanent magnetization (pDRM) based on anisotropic acquisition experiments of the isothermal remanent magnetization (IRM), we confirmed significantly shallow inclinations even for the flattening-corrected data set, implying northward transportation after the emplacement. Based on comparisons to expected paleomagnetic directions calculated from contemporaneous reference poles, we conclude that the allochthonous blocks, including south central Hokkaido, migrated northerly during the Early Cretaceous. Previous investigations of paleomagnetism and numerical modeling of burial processes of sedimentary basins indicate that some crustal blocks in Hokkaido and NE Japan experienced delayed transportation and eventually amalgamated with the mother continent by the end of the Paleogene.

Keywords: paleomagnetism, allochthonous block, lateral migration, Cretaceous, central Hokkaido

1. Introduction

The main part of the longitudinal mountainous range of Hokkaido is called the Sorachi-Yezo Belt (**Figure 1**). It is characterized by the occurrence of a Jurassic ophiolite (the Lower Sorachi Ophiolite) and a Cretaceous forearc basin sequence (the Nitarachi–Yezo Sequence). This ophiolite-forearc basin sequence is underlain tectonostratigraphically by Cretaceous accretionary complexes, whose eastern margin is defined as the Idonnappu Zone. Well-preserved accretion constituents have long been targets of geological surveys aiming to elucidate the evolutionary process of a longstanding arc-trench system (e.g., [1–3]). An implicit assumption in these studies was that the east-verging accretion had occurred somewhere along the northeastern Asian convergent margin and had a close genetic relationship with the autochthonous Cretaceous volcanic arc in the Sikhote Alin, Russia.

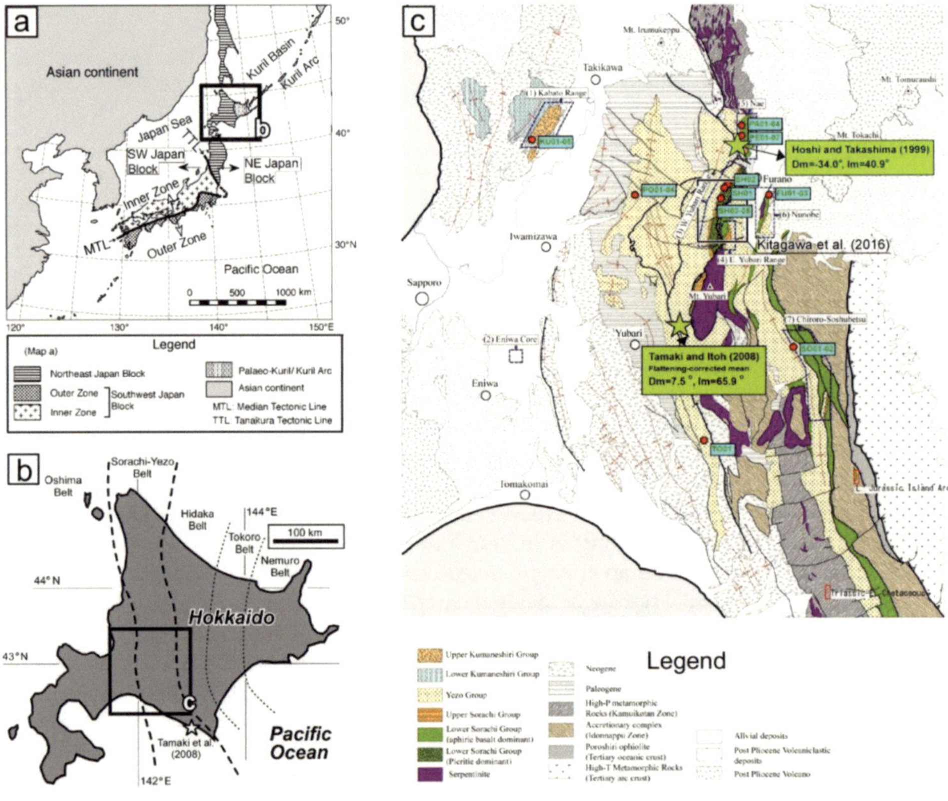

Figure 1. Index maps of study area. (a) Regional configuration and constituents of the Japanese Islands. (b) Pre-Cenozoic tectonic divisions in Hokkaido. (c) Geologic units around the sampling areas. Paleomagnetic sampling points are shown by dots. Stars and an enclosure indicate areas of previous paleomagnetic studies [4, 7, 10, 11].

However, recent paleomagnetic studies have cast doubt on part of the Mesozoic arc constituents having originated from a remote area. Tamaki et al. [4] found that the Upper Yezo Group (Campanian: [5]), distributed in the Urakawa area, preserves stable detrital remanent magnetization (DRM) characterized by a significantly shallow inclination. Compared to the expected

http://dx.doi.org/10.5772/67359

geomagnetic directions in northeastern Asia, untilted and flattening-corrected paleomagnetic directions require northward transportation of the Urakawa area by as much as 3400 km since the Cretaceous. The DRM of the Middle Yezo Group in the Oyubari area in central Hokkaido (**Figure 1c**; Cenomanian/Turonian: [6]) shows fairly deep inclinations, suggestive of an autochthonous origin [7], and the presumed Cretaceous Yezo forearc basin in East Asia (e.g., [8]) may be divided into some blocks with quite different tectonic histories.

To unravel this paradox, we executed paleomagnetic analyses on the Kumaneshiri, Sorachi, and Yezo Groups around the southern part of central Hokkaido (**Figure 1c**) in the course of the present study. Comparing the flattening-corrected inclination values obtained from the volcanic arc components and forearc sediments, we verified the self-consistency of the migration hypothesis of the ancient arc-trench system. The authors submit a quantitative constraint on tectonic models of the East Asian convergent margin.

2. Geological setting and sampling

Voluminous Late Jurassic to Early Cretaceous igneous and volcaniclastic rocks associated with tuffaceous sedimentary rocks are distributed in the Oshima and Sorachi-Yezo Belts of Hokkaido (**Figure 1b**) and are called the Kumaneshiri and Sorachi Groups, respectively. They both originated from subaqueous volcanism and show a strong resemblance in stratigraphic succession. **Figure 2** presents composite columnar sections of the late Mesozoic strata for selected survey areas. Takashima et al. [9] confirmed gravels of oolitic limestone and trachyandesite suffering subaerial oxidation in the lower part of the Nunobe Formation and deemed that part of the igneous rocks of the Sorachi Group was derived from island arcs. The presence of the oolitic limestone also implies sedimentation under a considerably warm climate.

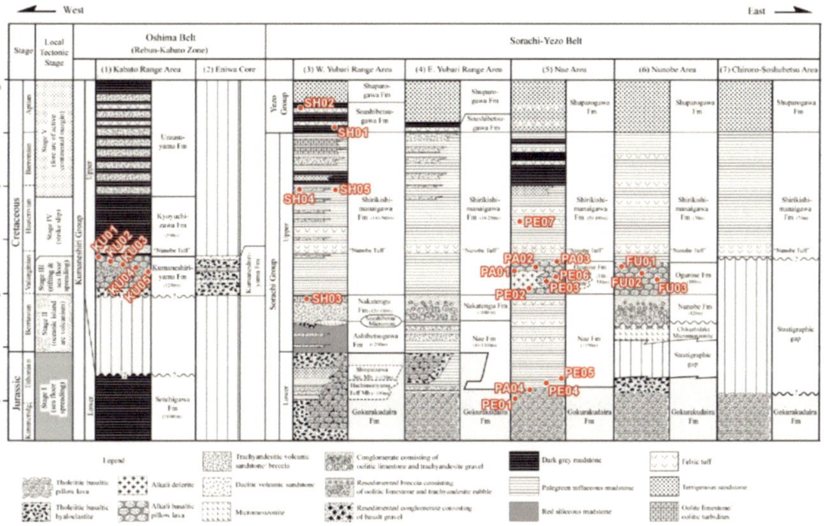

Figure 2. Composite columnar sections of the late Mesozoic strata for selected areas. See **Figure 1** for locations. Paleomagnetic sampling horizons are attached on the columns. The horizons of SO01, SO02, and TO01 are out of the range of the columns.

Because secondary magnetization caused by harsh alteration overprinted the magnetism, there are only a limited number of previous paleomagnetic works on the Sorachi Group. Hoshi and Takashima [10] measured remanent magnetization in dolerite and basalt (pillow lava) samples and obtained a shallow mean inclination implying northward transportation after emplacement. Large scatter in their data hindered precise discussion, but then Kitagawa et al. [11] analyzed limestone, volcaniclastic rock, and andesite with wider spatiotemporal coverage and reconfirmed significantly shallower inclinations than were expected for the coeval mother continent.

In this study, we obtained core samples to measure paleo- and rock magnetism at five sites from the Kumaneshiri Group (KU01–05), 17 sites from the Sorachi Group (FU01–03, PA01–04, PE01–07, SH03–05) and nine sites from the overlying Yezo Group (PO01–04, SH01–02, SO01–02, TO01) as shown in **Figures 1** and **2**. The lithofacies for all the sites are summarized in **Table 1**. Cores 25 mm in diameter were taken from each site using an engine drill, and the individual cores were oriented with a Brunton compass mounted on an aluminum orientation table. Along survey routes, sampling sites were selected to ensure that the structural attitudes needed for the tectonic tilt correction of paleomagnetic directions were clearly defined on outcrops. In the laboratory, cylindrical specimens 25 mm in diameter and 22 mm long were cut from each core sample.

Site	Lithology	Route
Yezo Group		
PO01	Silty tuff (weakly bioturbated)	Ponbetsu River (Nanashi Stream)
PO02	Medium sand-size tuff	Ponbetsu River (Nanashi Stream)
PO03	Medium – fine sand-size tuff	Ponbetsu River (Nanashi Stream)
PO04	Calcareous nodule	Ponbetsu River (Nanashi Stream)
SH01	Acidic tuff	Shirikishimanai River
SH02	Very fine sandstone	Shirikishimanai River
SO01	Tuffaceous mudstone	Soshubetsu River
SO02	Tuffaceous mudstone	Soshubetsu River
TO01	Acidic tuff	Hobetsu (Tomiuchi)
Kumaneshiri Group		
KU01	Pillow basalt	Kabato Range
KU02	Pillow basalt	Kabato Range
KU03	Basaltic volcaniclastic rock	Kabato Range
KU04	Basaltic volcaniclastic rock	Kabato Range
KU05	Basaltic volcaniclastic rock	Kabato Range
Sorachi Group		
FU01	Dolerite dike	Furano (Nunobe)
FU02	Pillow basalt	Furano (Nunobe)

Site	Lithology	Route
FU03	Pillow basalt	Furano (Nunobe)
PA01	Pillow basalt	Panketeshimanai River
PA02	Dolerite dike	Panketeshimanai River
PA03	Pillow basalt	Panketeshimanai River
PA04	Pillow basalt	Panketeshimanai River
PE01	Pillow basalt	Penketeshimanai River
PE02	Dolerite sill	Penketeshimanai River
PE03	Pillow basalt	Penketeshimanai River
PE04	Volcanic sandstone	Penketeshimanai River
PE05	Volcanic sandstone	Penketeshimanai River
PE06	Mudstone	Penketeshimanai River
PE07	Acidic tuff	Penketeshimanai River
SH03	Oolitic limestone (turbidite)	Shirikishimanai River
SH04	Volcanic fine sandstone containing mud patch	Shirikishimanai River
SH05	Volcanic mudstone	Shirikishimanai River

Table 1. Description of paleomagnetic samples.

3. Paleomagnetism

3.1. Basic methods

Bulk initial magnetic susceptibility was measured for all the specimens using a Bartington susceptibility meter (MS2). We conducted progressive thermal demagnetization (PThD) tests on selected pilot specimens from each site, except on samples that were too fragile for repeated heating, for which progressive alternating-field demagnetization (PAFD) testing was adopted. Natural remanent magnetization (NRM) was measured using a cryogenic magnetometer (760-R SRM, 2-G Enterprises) in a magnetically shielded room at Kyoto University and spinner magnetometers (SMM-85, Natsuhara-Giken; SSM-1A, Schonstedt Instrument) at Osaka Prefecture University. The PThD test was performed up to 680°C in air using a noninductively wound electric furnace with an internal residual magnetic field of less than 10 nT. The PAFD test was carried out stepwise up to 100 mT with a three-axis tumbler contained in μ-metal shield envelopes.

3.2. Demagnetization tests

Figure 3 depicts typical results of the PThD testing showing in situ coordinates. Focusing on the Pinneshiri area in the Kabato Range, for five sites in the Kumaneshiri Group (KU01–05), we found the stable components to have a converging trend on origin of the vector-demagnetization diagrams across a broad distribution of unblocking temperatures (T_{UB}) up to 600°C after the northerly component was demagnetized at around 300°C. In the eastern part of the sampling region (the Nokanan area), we successfully isolated similar high-T_{UB} components of remanent magnetization for five sites in the Sorachi Group (PE04, PE07, and SH03–05) and two sites in the lower Yezo Group (SH01–02). The directions of the characteristic remanent magnetization (ChRM) were calculated using a three-dimensional least squares analysis technique after [12].

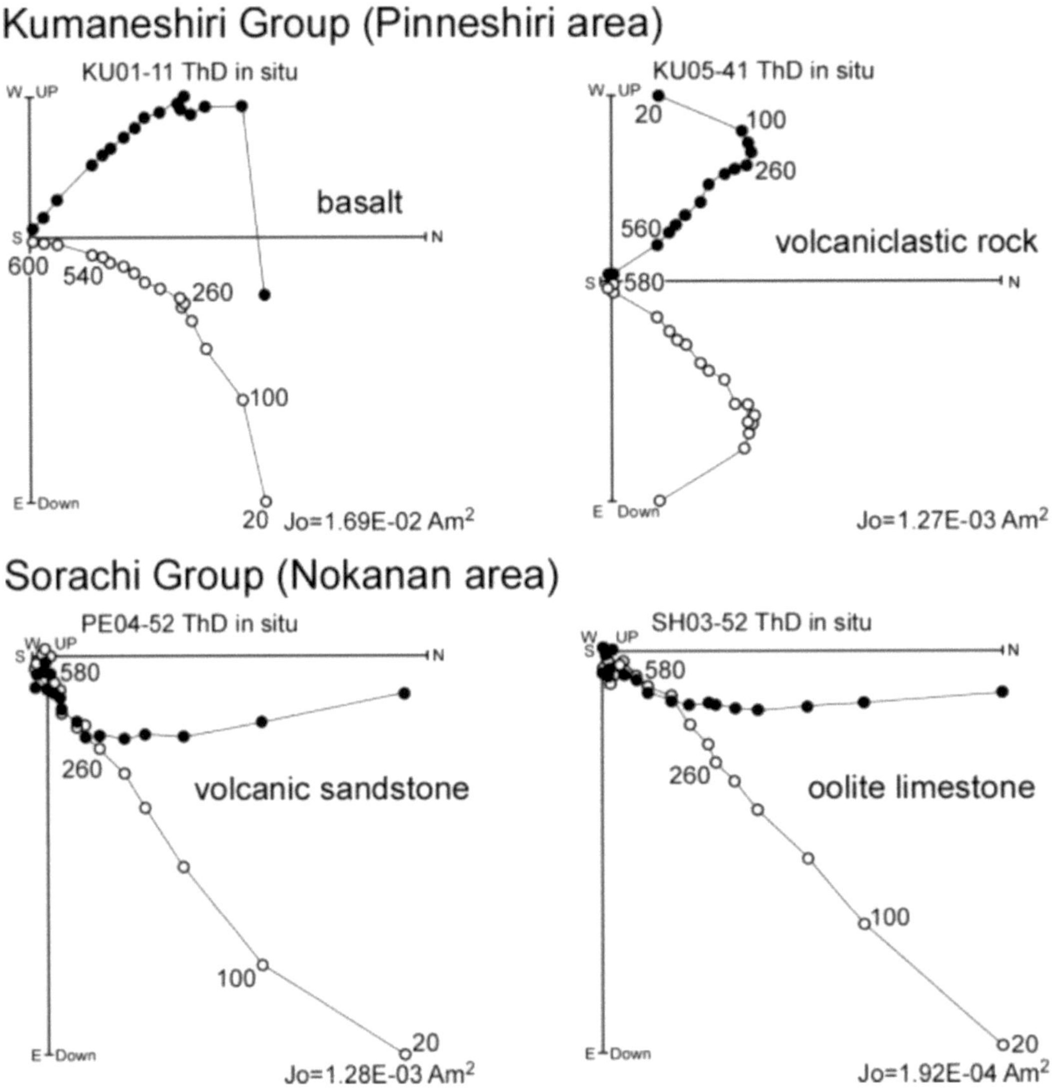

Figure 3. Typical results of progressive thermal demagnetization (PThD) in in situ coordinates. On the vector-demagnetization diagrams, solid (open) circles are projections of vector end-points on a horizontal (N-S vertical) plane. Numbers are demagnetization levels in °C.

Five site-mean ChRM directions obtained from the Kumaneshiri Group show normal polarity, and the precision parameter (κ) improves after tilt correction (**Figure 4**). Seven sites in the

Sorachi and Yezo Groups in the Nokanan area are clustered into antipodal ChRM directions (**Figure 4**), which after polarity inversion, pass a positive reversal test at a 95% confidence level ($f = 1.04 < Fc$) after [13]. Although more rigorous verification may be desirable, we tentatively regard the two data sets as primary records of the earth's dipole field and utilize them for tectonic discussion.

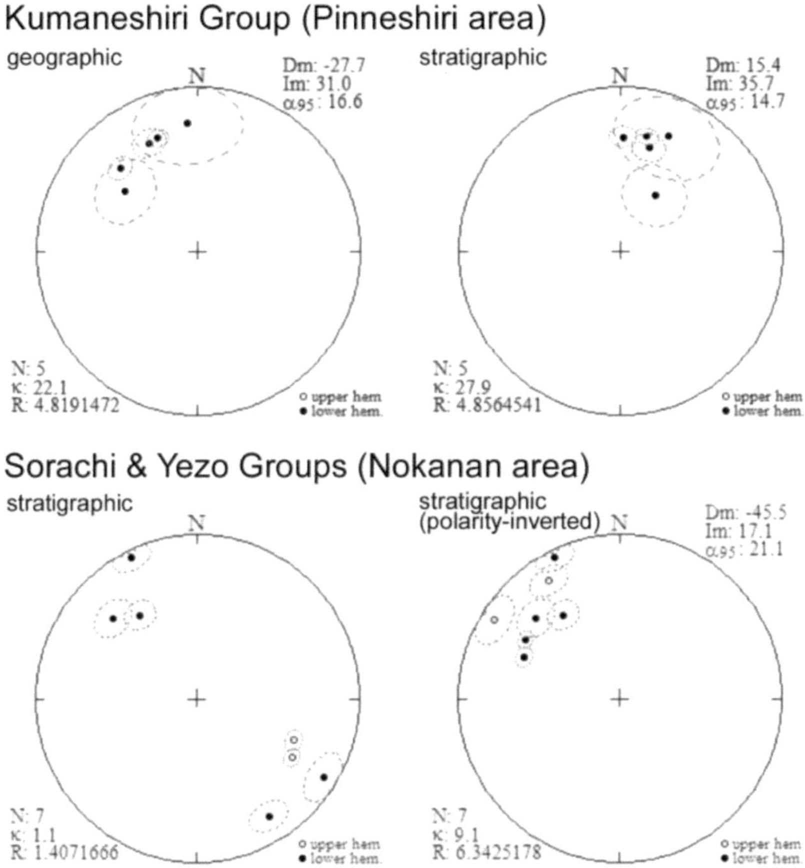

Figure 4. Site-mean directions of high T_{UB} components of the Kumaneshiri, Sorachi, and Yezo Groups in geographic and stratigraphic coordinates on equal-area projections. Solid (open) symbols are on the lower (upper) hemisphere of the equal-area projections. Dotted ovals are 95% confidence limits of site-means.

3.3. Identification of ferromagnetic minerals

3.3.1. Spectrum of coercive force (H_C)

In order to identify carriers of magnetic components in the samples, we undertook isothermal remanent magnetization (IRM) experiments. Stepwise IRM acquisition was performed according to an analytical technique developed by Kruiver et al. [14]. **Figure 5** shows the linear acquisition plot (LAP) and gradient of acquisition plot (GAP) of the IRM acquired in direct magnetic fields of up to around 3 T. As shown by these examples, the plots generated from a majority of the IRM data can be matched by single magnetic components with relatively low $B_{1/2}$ values (the field at which half the IRM saturation is reached), indicating the existence of low H_C ferromagnetic minerals. On the basis of the T_{UB} spectra mentioned before, we believe the remanent magnetization of the major samples resides in titanomagnetite.

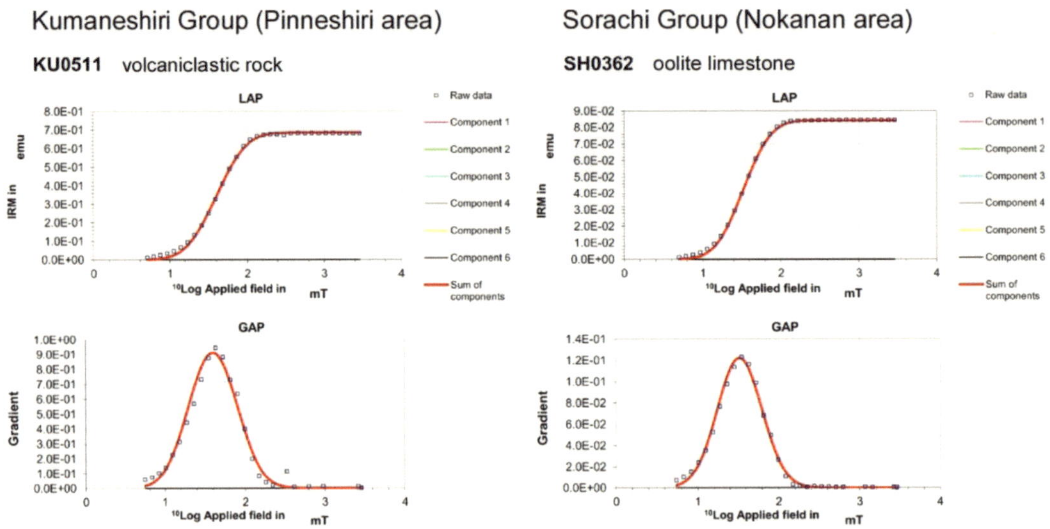

Figure 5. Linear acquisition plot (LAP) and gradient of acquisition plot (GAP) of the isothermal remanent magnetization (IRM) acquired in direct magnetic fields of up to around 3 T. Specimens were processed after alternating field demagnetization at 100 mT.

3.3.2. Thermal demagnetization of orthogonal IRMs

We performed PThD of composite IRMs on selected specimens. Based upon the procedure proposed by Lowrie [15], composite IRMs were imparted by applying direct magnetic fields of 3.0, 0.4 and then 0.12 T to the specimens in three orthogonal directions. As shown in **Figure 6**, the decay curve of the IRM components derived from PThD testing indicates that the dominant magnetic phase is generally the low H_C (<0.12 T) soft fraction with a broad spectrum of T_{UB} up to 580°C. In such a case, the major carrier of the high-T_{UB} component of the NRM is titanomagnetite. Smaller amounts of medium ($0.12 < H_C < 0.4$ T) and hard ($0.4 < H_C < 3.0$ T) fractions were identified. As for the basaltic rocks of the Kumaneshiri Group (KU0141, KU0231, and KU0511), a minor medium fraction is interpreted to be carried by fine (SD-size) grains of magnetite because they have T_{UB} spectra up to 580°C. A small amount of the hard fraction in volcanic samples (PE0421 in the Sorachi Group, SH0111 in the Yezo Group) is carried by hematite because they have T_{UB} spectra up to 680°C. These experiments clarified that the single-component NRMs preserved in samples of the Kumaneshiri, Sorachi and Yezo Groups are carried by a mixture of titanomagnetite and hematite mixed in various ratios.

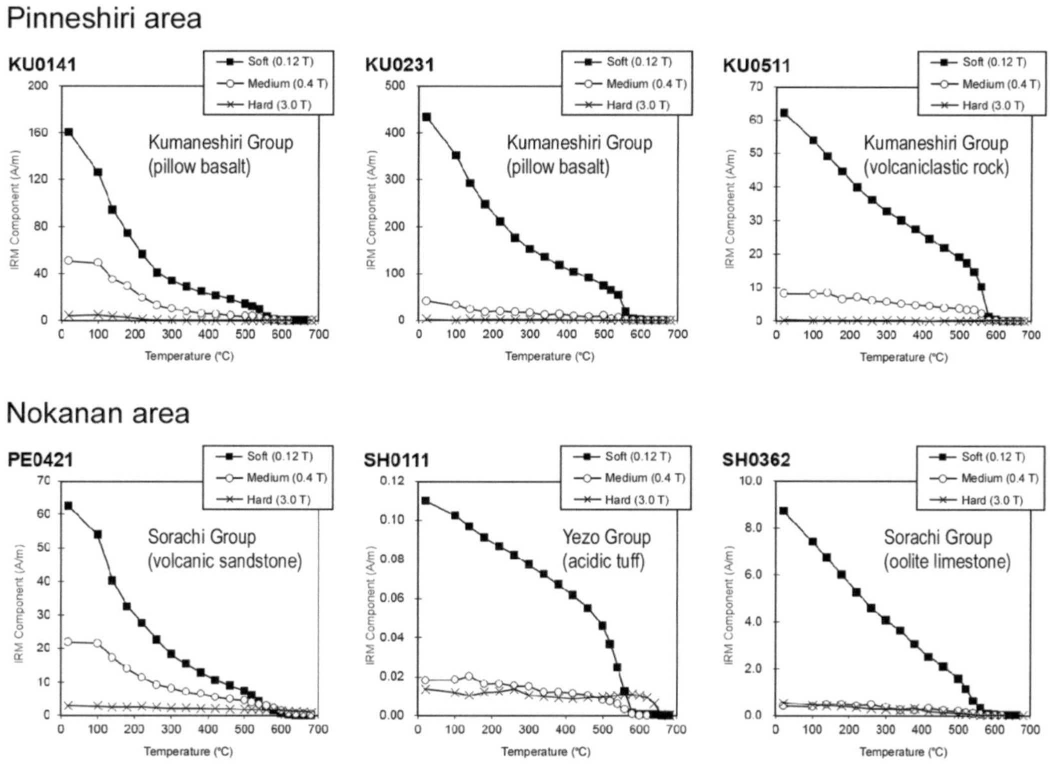

Figure 6. Thermal demagnetization curves of orthogonal IRMs for representative samples.

3.4. Magnetic fabric

3.4.1. Anisotropy of magnetic susceptibility

In order to describe the magnetic fabric of the samples, we determined each specimen's anisotropy of magnetic susceptibility (AMS). Measurement was done using a KappaBridge KLY-3S magnetic susceptibility meter (AGICO). All the results of AMS measurements are presented graphically in **Figure 7** except for site KU01, for which there were too little data to obtain statistical parameters. The tilt-corrected AMS fabric (principal susceptibility axes) presented on an equal-area projection for the volcanic and volcaniclastic rocks (e.g., FU01–03, KU02–05, PA01–04, PE01–03, SH01) seems to be irrelevant to the sedimentary surface (horizontal plane). On the other hand, a majority of the sedimentary rocks (e.g., PE04–06, PO01–04, SH02, SO01–02) exhibit arrangements of AMS axes bound to the bedding plane, namely, the minimum axis (K_3) is nearly perpendicular to the bedding in stratigraphic coordinates, which suggests that the samples preserve the original sedimentary structure without significant tectonic distortion, and the *T* parameter has positive values indicative of oblate fabric. Because the microfabrics being bound to the bedding plane may introduce shallowing of inclinations of the post-depositional detrital remanent magnetization (pDRM), we tested the degree of remanence anisotropy as described in the next section.

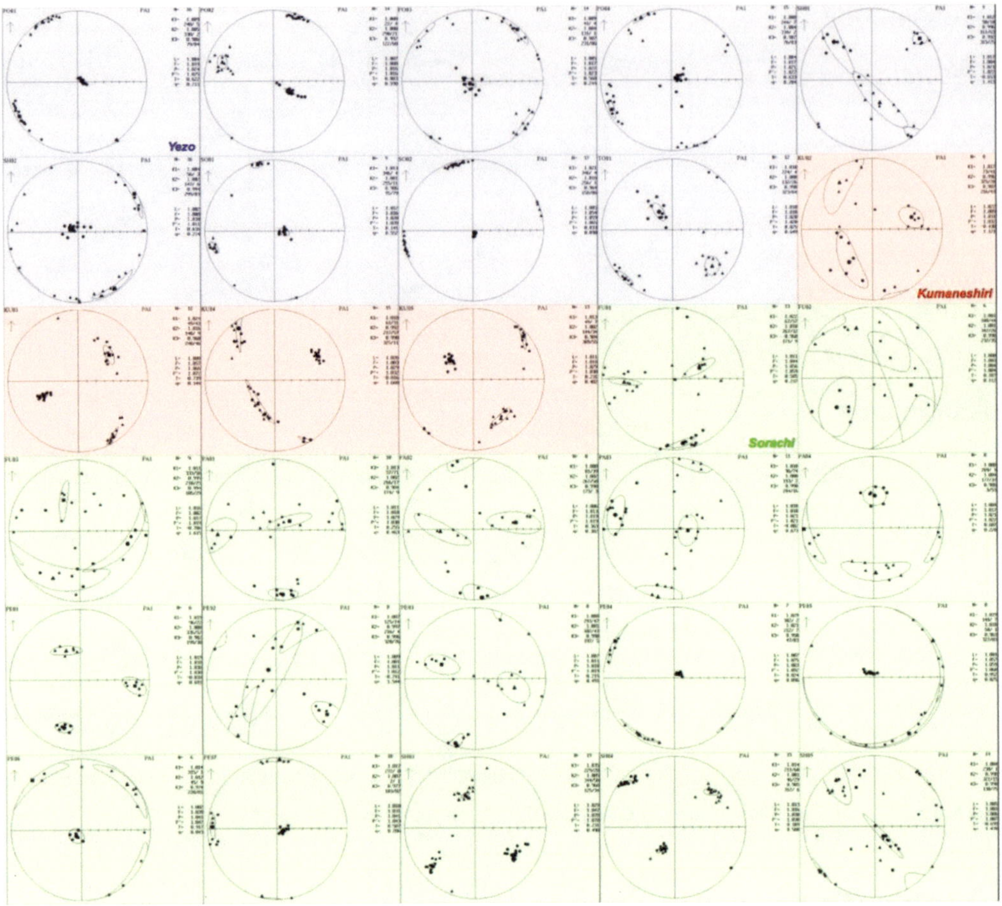

Figure 7. Anisotropy of magnetic susceptibility (AMS) fabric (principal susceptibility axes) for all specimens from each site of the Kumaneshiri, Sorachi and Yezo Groups plotted on the lower hemisphere of an equal-area projection. Data are shown in stratigraphic coordinates. Square, triangular and circular symbols represent orthogonal maximum (K_1), intermediate (K_2), and minimum (K_3) AMS principal axes, respectively, and larger symbols show their mean directions. Ovals surrounding the mean directions of the three axes are 95% confidence regions based upon Bingham statistics.

3.4.2. Anisotropy in IRM acquisition

We performed inclination shallowing testing based on the method of [16] using IRM anisotropy. On one specimen per site, from which we obtained the ChRM, we applied a direct magnetic field at 45° to the bedding plane to avoid any field impressed anisotropy [17]. We then measured the IRM, which was parallel (IRMx) and perpendicular (IRMz) to the bedding. **Figure 8** shows typical IRM acquisition curves for IRMz and IRMx. The value of IRMz is lower than IRMx for the entire range of acquisition (**Figure 8a**), suggesting that ferromagnetic minerals carrying NRM are anisotropic and follow a similar trend. The ratio IRMz/IRMx can be uniquely related to the amount of inclination shallowing (tan I/tan I_F = IRMz/IRMx; I_F = inclination of the field in which remanence was acquired). The average IRMz/IRMx ratio for each sample was determined from the best-fit slope of IRMz against IRMx (**Figure 8b**).

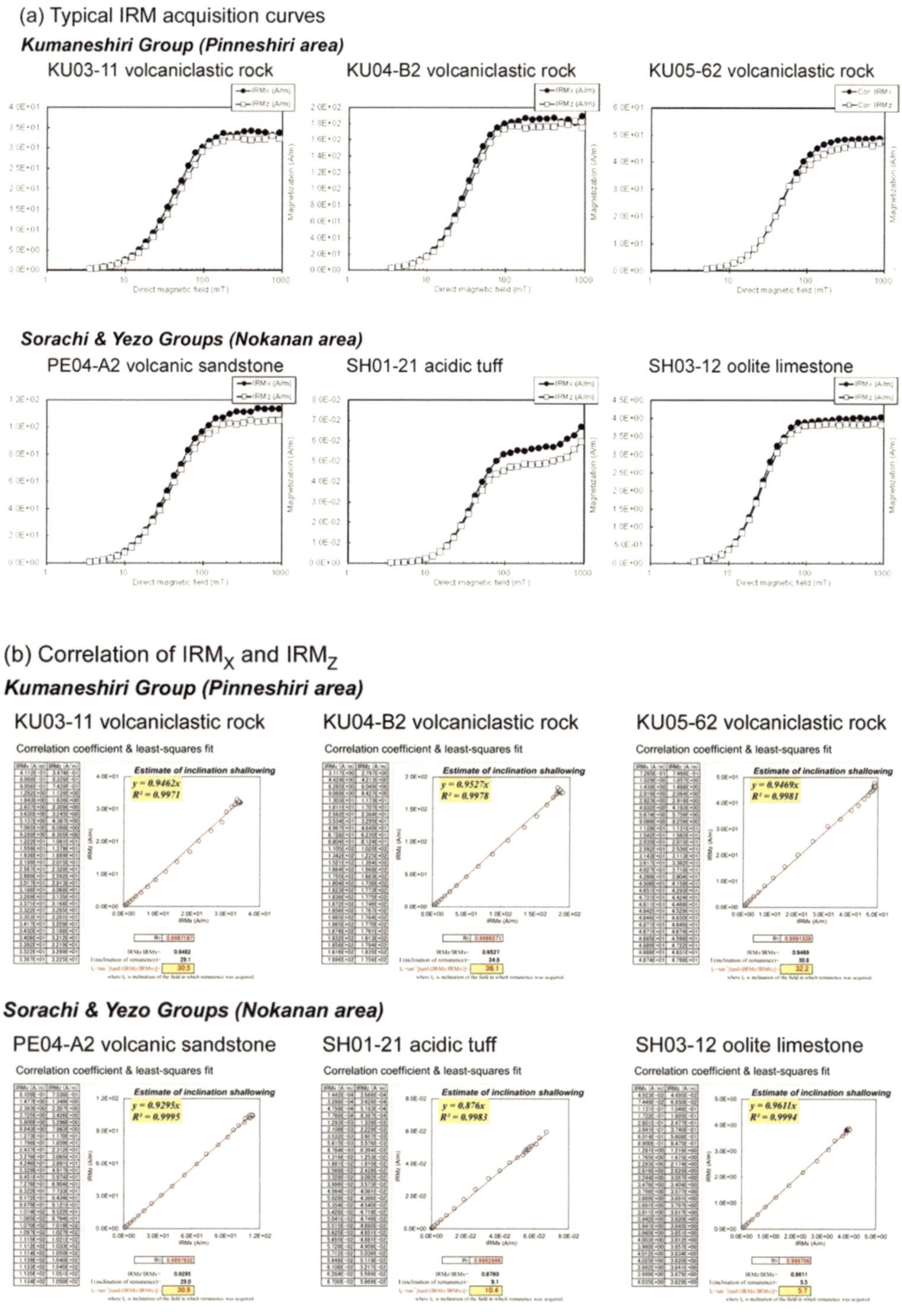

Figure 8. (a) Typical IRM acquisition curves for the bedding-normal (IRMz) and bedding-parallel components (IRMx) for selected samples. (b) Gradient of the best-fit correlation line of IRMz versus IRMx used to determine the IRMz/IRMx ratio, which gives an estimate of inclination shallowing in the sediments.

4. Discussion

4.1. Implications of shallowing correction

Inclination shallowing was found for most of the analyzed samples. As the shallowing occurred during postdepositional compaction prior to tectonic tilting, we applied the shallowing correction for the untilted data set in the study area. **Figure 9** presents the primary magnetic directions of the Kumaneshiri, Sorachi, and Yezo Groups before and after shallowing correction. As suggested by Jackson et al. [18], anisotropy of IRM or anhysteretic remanent magnetization (ARM) may decrease as a result of postdepositional processes such as electrostatic and coagulation effects, whereas the direction of the primarily acquired pDRM is immune from such secondary effects. In that case, our flattening estimation, based on IRM anisotropy, may underestimate the actual amount of inclination flattening. Our corrected data set (**Table 2**) would then provide a minimum estimate of the paleolatitude. Based on the assumption that the data sets are records of the earth's dipole magnetic field, the inclinations of formation-means for the Pinneshiri and Nokanan areas correspond to paleolatitudes of 20°N and 9°N, respectively.

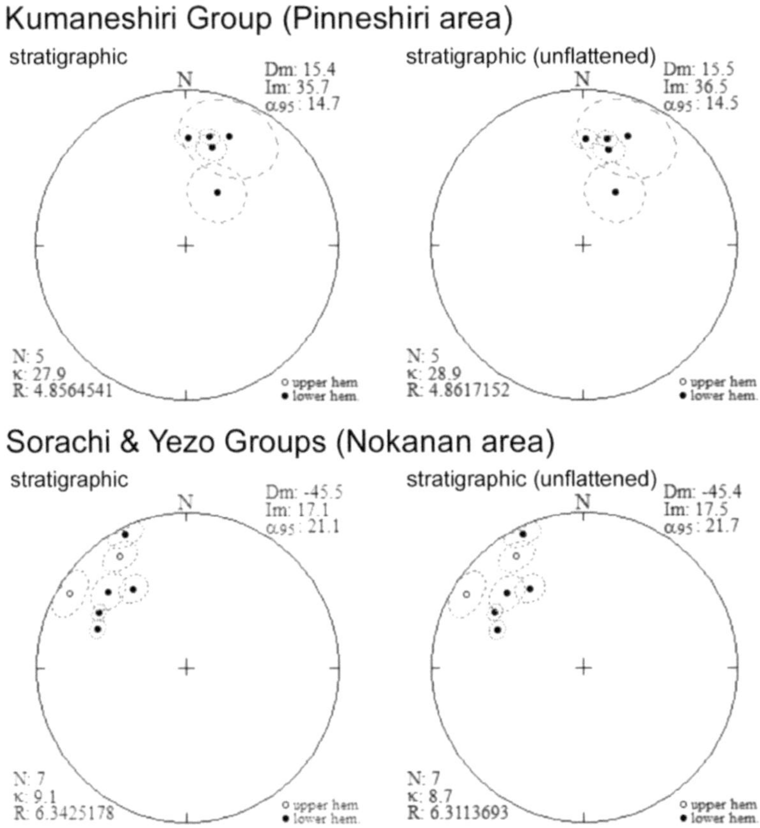

Figure 9. The primary magnetic directions of the Kumaneshiri, Sorachi and Yezo Groups before and after inclination shallowing correction in stratigraphic coordinates on equal-area projections. Solid (open) symbols are on the lower (upper) hemisphere of the equal-area projections. Dotted ovals are 95% confidence limits of site-means.

http://dx.doi.org/10.5772/67359

Site	N	D (°)	I (°)	D_C (°)	I_C (°)	I_F (°)	α_{95} (°)	κ
Yezo Group								
SH01	8	−174.7	−71.4	121.7	9.1	10.4	10.3	26.4
SH02	9	−116.9	−17.7	113.1	−35.6	−38.1	4.7	121.2
Kumaneshiri Group								
KU01	9	−4.8	22.5	22.3	25.0	25.0	23.2	5.9
KU02	7	−51.0	41.1	31.4	57.4	57.4	15.7	15.7
KU03	7	−19.7	27.6	12.9	29.1	30.5	4.8	158.5
KU04	7	−24.4	28.1	15.0	34.8	36.1	7.7	62.6
KU05	7	−43.4	30.5	1.0	30.8	32.2	5.9	106.6
Sorachi Group								
PE04	7	54.6	49.8	−46.7	29.0	30.8	9.3	43.5
PE07	7	−146.5	−57.9	121.5	−31.9	−31.9	4.2	158.7
SH03	11	40.6	39.9	−25.2	5.5	5.7	7.7	36.4
SH04	8	67.2	29.4	−34.7	37.7	37.7	7.6	53.4
SH05	8	−172.0	−49.3	148.3	16.5	16.7	8.4	34.7

N is number of specimens; D and I are in situ site-mean declination and inclination, respectively; D_C and I_C are untilted site-mean declination and inclination, respectively; I_F is inclination of the field in which remanence was acquired; α_{95} is the radius of 95% confidence circle; κ is Fisher's precision parameter.

Table 2. Mesozoic paleomagnetic directions.

4.2. Possible allochthonous blocks in central Hokkaido

The present study ratified the hypothetical transportation from low latitudes proposed by the authors in Refs. [10, 11]. The central part of Hokkaido, however, does not seem to have migrated en bloc because the Mesozoic paleomagnetic records in some areas are indicative of autochthonous origin. Our PThD examinations for the Yezo Group distributed along the Ponbetsu River (PO01–04) imply deep inclinations, although they were not magnetically stable enough to determine site-mean ChRM directions. As with the Oyubari area studied by Tamaki and Itoh [7] (see **Figure 1**), we posit that the western wing of the N-S trending anticline may be composed of in situ blocks, whereas we obtained significantly shallow inclinations from the core and eastern wing of the structure. An unsolved problem is that a clear geologic boundary between the blocks with mixed origins has never been detected.

Dismemberment of amalgamated blocks during the late Cenozoic collision between the Kurile and northeastern Japan arcs is also a knotty problem for reliable paleoreconstruction. Exploration drilling was executed in 1997 in central Hokkaido (42.9943°N, 142.0228°E; [19]). The vertically drilled borehole reached 4465 m depth and confirmed that the Cretaceous to Tertiary strata was repeatedly stacked by remarkable west-vergent thrusts. Actually, the western foothills of the backbone mountains consist of multiple-stacked thrust horses and the flat-lying horst-graben of the Paleogene and Cretaceous igneous basement [20]. Based on an interpretation of regional seismic profiles, Kazuka et al. [21] estimated the east-west crustal shortening across the Hidaka Mountains to be ~60 km. Thus, the origin of the allochthonous terranes should be clarified through further investigation of the three-dimensional structure of the island.

4.3. Estimate of N-S transportation

In order to determine the amount of tectonic movement of geologic units now distributed in south central Hokkaido, the expected direction should be calculated from the contemporaneous reference pole of the North China Block (NCB) after [22]. Because it formed a single entity with Siberia and Mongolia around the Early Cretaceous [23], the whole block can represent a coherent part of East Asia since the Cretaceous time. **Figure 10** presents a summary of Mesozoic to Cenozoic paleomagnetic information around the study area. Our results basically agree with the estimate of N-S transportation by the authors in Refs. [10, 11]. Based on temporal decreases in positive F values defined by Beck [24], allochthonous blocks migrated northerly during the Early Cretaceous. Paleomagnetic data reported from the Oyubari area in central Hokkaido [7] show an affinity to the NCB data. Considering a reconstructed subduction history by Ueda and Miyashita [25], simultaneous events of amalgamation on the continental margin may have occurred from 100 to 90 Ma.

Kimura et al. [26] regarded the Sorachi Group as a constituent of an enormous oceanic plateau that was driven to collide against the continent by rapid northward movement of the Izanagi Plate [27]. It seems, however, that their hypothesis clashes with the evidence of arc volcanism, as mentioned above. Takashima et al. [28] proposed an alternative idea that the geologic unit was transported by left-lateral slips on the margin between the Eurasian and Izanagi Plates. Although oblique subduction is a plausible cause of the large migration, lateral motion on the plate margin changed to a dextral direction after the demise of the Izanagi Plate [29]. In order to reconcile such controversial points, we present a comprehensive paleoreconstruction in the last chapter of this book.

Paleomagnetic analysis made it clear that some crustal blocks experienced delayed transportation. A composite mean of the Upper Yezo Group in the Urakawa area [4] gives a paleolatitude of 16.7°N. Using the NCB expected direction for comparison, northward transportation since the Late Cretaceous appears to be 3400 km. Based on geochemical modeling, Itoh et al. [29] suggested continued subsidence of the allochthonous 'forearc' region containing Urakawa area through the Paleogene. The considerable thickness of the missing unit is not attributed to eustatic sea-level changes but to tectonic subsidence of the forearc, which implies that a subduction erosion process [30] was active. It is noteworthy that an autochthonous block in central Hokkaido also suffered Paleogene subsidence. Tamaki et al. [31] executed 1D basin modeling on the basis of organic maturation data obtained from a deep borehole (MITI Yubari; [19]). Their burial history was better constrained because the data set contained maturity levels and the present thickness of the Eocene sedimentary units. They found accelerated accumulation rates during the Paleogene, which is an indicator of the emergence of a foreland basin setting (e.g., [32]). Simultaneous inversion of these areas implies that an amalgamation of migrated terranes occurred by the end of the Paleogene.

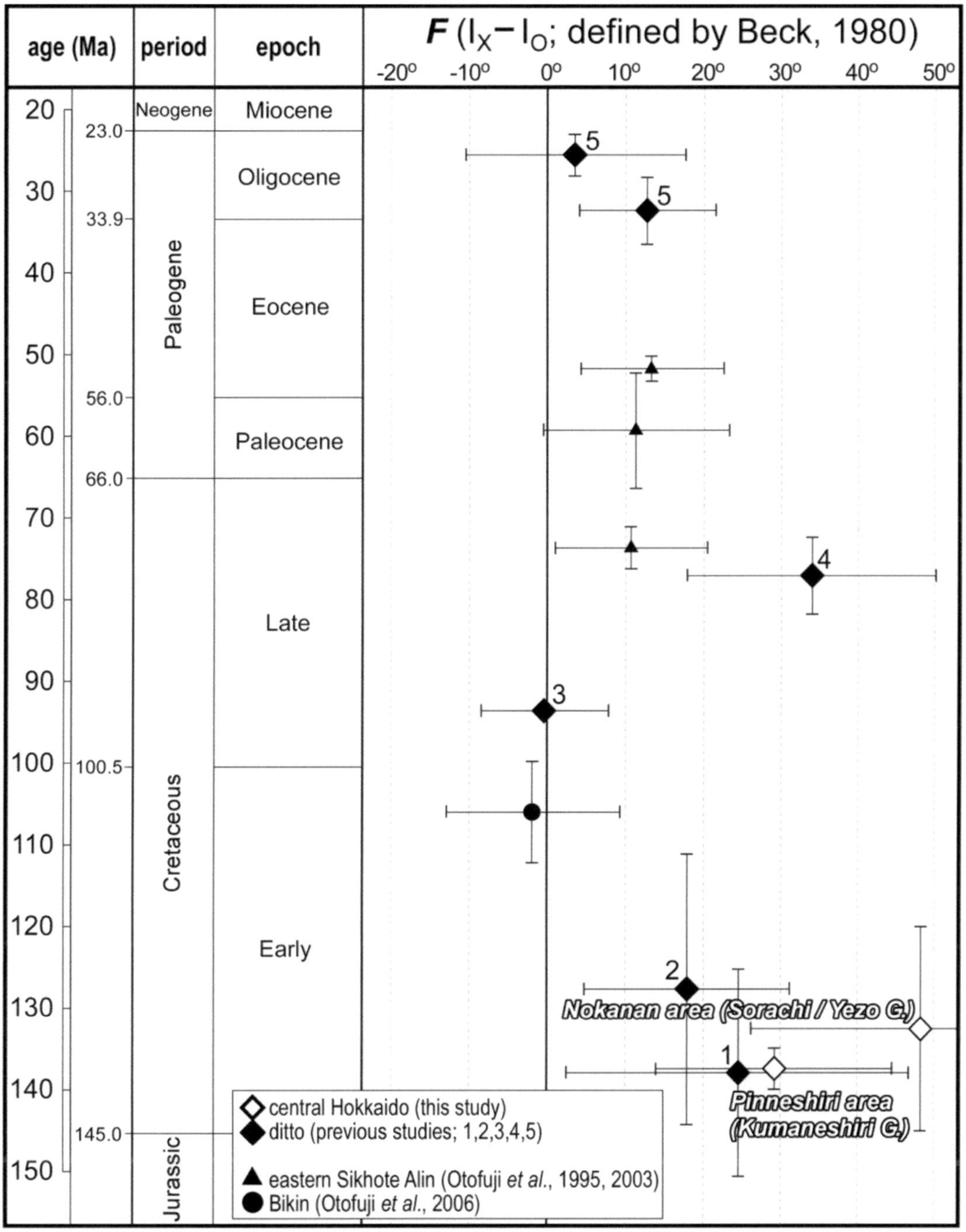

Figure 10. Plot of inclination flattening (F) (F = I [expected] – I [observed]; [24]) versus age for central Hokkaido (1 = [10]; 2 = [11]; 3 = [7]; 4 = [4]; 5 = [33]), eastern Sikhote Alin and Bikin. Stratigraphic positions of data in eastern Sikhote Alin and Bikin are after Otofuji et al. [34, 35] and Otofuji et al. [36], respectively. The age range of the Oyubari data (3: [7]) is shorter than the height of the median symbol.

5. Conclusions

Lateral migration of the Oshima and Sorachi-Yezo Belts within south central Hokkaido was quantitatively evaluated by means of paleomagnetic analyses. We tested the remanence stability of the Late Jurassic to Early Cretaceous voluminous igneous succession of the Kumaneshiri and Sorachi Groups and the overlying forearc sediments of the Cretaceous Yezo Group through rock magnetic experiments. Twelve of the sites yielded characteristic primary components residing in mixtures of titanomagnetite and hematite having various mixing ratios. Although anisotropic acquisition of the isothermal remanent magnetization suggested shallowing of inclinations of the post-depositional detrital remanent magnetization, we confirmed significantly shallow inclinations even for the flattening-corrected data, implying northward transportation after emplacement. Based on comparisons to expected paleomagnetic directions calculated from contemporaneous reference poles, we concluded that the allochthonous blocks, including central Hokkaido, migrated northerly during the Early Cretaceous. Previous research studies of paleomagnetism and numerical basin modeling of burial processes indicate that some crustal blocks experienced delayed transportation and eventually amalgamated with the mother continent by the end of the Paleogene.

Acknowledgements

The authors extend their appreciation to Naoto Ishikawa for use of the rock magnetic laboratory at Kyoto University. We thank Yuki Bito and Takashi Inoue for their earnest help during field work and experiments.

Author details

Yasuto Itoh[1]* and Reishi Takashima[2]

*Address all correspondence to: yasutokov@yahoo.co.jp

1 Graduate School of Science, Osaka Prefecture University, Osaka, Japan

2 The Center for Academic Resources and Archives (Tohoku University Museum), Tohoku University, Sendai, Japan

References

[1] Niida K, Kito N. Cretaceous arc-trench system in Hokkaido. Monograph of the Association for the Geological Collaboration in Japan. 1986; 31: 379–402.

[2] Kiyokawa S. Geology of the Idonnappu Belt, central Hokkaido, Japan: evolution of a Cretaceous accretionary complex. Tectonics. 1992; 11: 1180–1206.

[3] Ueda H, Kawamura M, Niida K. Accretion and tectonic erosion processes revealed by the mode of occurrence and geochemistry of greenstones in the Cretaceous accretionary complexes of the Idonnappu Zone, southern central Hokkaido, Japan. The Island Arc. 2000; 9: 237–257.

[4] Tamaki M, Oshimbe S, Itoh Y. A large latitudinal displacement of a part of Cretaceous forearc basin in Hokkaido, Japan: paleomagnetism of the Yezo Supergroup in the Urakawa area. Journal of the Geological Society of Japan. 2008; 114: 207–217.

[5] Sakai A, Kanie Y. Geology of the Nishicha District, with Geological Sheet Map at 1:50,000. Tsukuba: Geological Survey of Japan; 1986. 92 p.

[6] Takashima R, Kawabe F, Nishi H, Moriya K, Wani R, Ando H. Geology and stratigraphy of forearc basin sediments in Hokkaido, Japan: Cretaceous environmental events on the north-west Pacific margin. Cretaceous Research. 2004; 25: 365–390.

[7] Tamaki M, Itoh Y. Tectonic implications of paleomagnetic data from upper Cretaceous sediments in the Oyubari area, central Hokkaido, Japan. Island Arc. 2008; 17: 270–284.

[8] Ando H. Stratigraphic correlation of Upper Cretaceous to Paleocene forearc basin sediments in Northeast Japan: cyclic sedimentation and basin evolution. Journal of Asian Earth Sciences. 2003; 21: 921–935.

[9] Takashima R, Miyamoto Y, Nishi H, Yoshida T. Geology and stratigraphy of the Sorachi and Yezo Groups in the Tokyo University Forests in Hokkaido, Japan. Bulletin of Tokyo University Forests. 2002; 108: 57–76.

[10] Hoshi H, Takashima R. Paleomagnetic analysis for some volcanic rocks of the Sorachi Group in the Furano area, central Hokkaido, Japan. Bulletin of the Mikasa City Museum, Natural Science. 1999; 3: 23–30.

[11] Kitagawa Y, Takashima R, Itoh Y. Paleomagnetism of the Sorachi and Yezo Group in the Ashibetsu area, central Hokkaido, Japan. Bulletin of the Tohoku University Museum. 2016; 15: 109–125.

[12] Kirschvink JL. The least-squares line and plane and the analysis of palaeomagnetic data. Geophysical Journal of the Royal Astronomical Society. 1980; 62: 699–718.

[13] McFadden PL, Jones DL. The fold test in palaeomagnetism. Geophysical Journal of the Royal Astronomical Society. 1981; 67: 53–58.

[14] Kruiver PP, Dekkers MJ, Heslop D. Quantification of magnetic coercivity components by the analysis of acquisition curves of isothermal remanent magnetisation. Earth and Planetary Science Letters. 2001; 189: 269–276.

[15] Lowrie W. Identification of ferromagnetic minerals in a rock by coercivity and unblocking temperature properties. Geophysical Research Letters. 1990; 17: 159–162.

[16] Hodych JP, Buchan KL. Early Silurian palaeolatitude of the Springdale Group redbeds of central Newfoundland: a palaeomagnetic determination with a remanence anisotropy test for inclination error. Geophysical Journal International. 1994; 117: 640–652.

[17] Tauxe L, Constable C, Stokking L, Badgley C. Use of anisotropy to determine the origin of characteristic remanence in the Siwalik Red Beds of northern Pakistan. Journal of Geophysical Research. 1990; 95: 4391–4404.

[18] Jackson MJ, Banerjee SK, Marvin JA, Lu R, Gruber W. Detrital remanence, inclination errors, and anhysteretic remanence anisotropy: quantitative model and experimental results. Geophysical Journal International. 1991; 104: 95–103.

[19] JNOC (Japan National Oil Corporation). Report for the Geological Study of MITI Yubari Borehole, 1997 Fiscal Year. Tokyo: Japan National Oil Corporation; 1998.

[20] Itoh Y, Ishiyama T, Nagasaki Y. Deformation mode in the frontal edge of an arc-arc collision zone: subsurface geology, active faults and paleomagnetism in southern central Hokkaido, Japan. Tectonophysics. 2005; 395: 81–97.

[21] Kazuka T, Kikuchi S, Ito T. Structure of the foreland fold-and-thrust belt, Hidaka Collision Zone, Hokkaido, Japan: re-processing and re-interpretation of the JNOC seismic reflection profiles 'Hidaka' (H91-2 and H91-3). Bulletin of Earthquake Research Institute. 2002; 77: 97–109.

[22] Gilder S, Courtillot V. Timing of the North-South China collision from new middle to late Mesozoic paleomagnetic data from the North China block. Journal of Geophysical Research. 1997; 102: 17713–17727.

[23] Hankard F, Cogne J-P, Quidelleur X, Bayasgalan A, Lkhagvadorj P. Palaeomagnetism and K-Ar dating of Cretaceous basalts from Mongolia. Geophysical Journal International. 2007; 169: 898–908.

[24] Beck ME Jr. Paleomagnetic record of plate-margin tectonic processes along the western edge of North America. Journal of Geophysical Research. 1980; 85: 7115–7131.

[25] Ueda H, Miyashita S. Tectonic accretion of a subducted intraoceanic remnant arc in Cretaceous Hokkaido, Japan, and implications for evolution of the Pacific northwest. The Island Arc. 2005; 14: 582–598.

[26] Kimura G, Sakakibara M, Okamura M. Plumes in central Panthalassa? Deductions from accreted oceanic fragments in Japan. Tectonics. 1994; 13: 905–916.

[27] Engebretson DC, Cox A, Gordon RC. Relative motions between oceanic and continental plates in the Pacific Basin. Geological Society of America Special Paper. 1985; 206: 1–59.

[28] Takashima R, Nishi H, Yoshida T. Late Jurassic-Early Cretaceous intra-arc sedimentation and volcanism linked to plate motion change in northern Japan. Cambridge: Geological Magazine, Cambridge University Press; 2006. doi: 10.1017/S001675680600255X

[29] Itoh Y, Takano O, Kusumoto S, Tamaki M. Mechanism of long-standing Cenozoic basin formation in central Hokkaido: an integrated basin study on an oblique convergent margin. Progress in Earth and Planetary Science. 2014; 1: 6. doi: 10.1186/2197-4284-1-6

[30] von Huene R, Lallemand S. Tectonic erosion along the Japan and Peru convergent margins. Geological Society of America Bulletin. 1990; 102: 704–720.

[31] Tamaki M, Tsuchida K, Itoh Y. Geochemical modeling of sedimentary rocks in the central Hokkaido, Japan: episodic deformation and subsequent confined basin-formation along the eastern Eurasian margin since the Cretaceous. Journal of Asian Earth Sciences. 2009; 34: 198–208.

[32] Allen PA, Allen JR. Basin Analysis: Principles and Applications, second ed. Oxford: Blackwell Publishing; 2005. 549 p.

[33] Tamaki M, Kusumoto S, Itoh Y. Formation and deformation processes of late Paleogene sedimentary basins in southern central Hokkaido, Japan: paleomagnetic and numerical modeling approach. Island Arc. 2010; 19: 243–258.

[34] Otofuji Y, Matsuda T, Itaya T, Shibata T, Matsumoto M, Yamamoto T, Morimoto C, Kulinich RG, Zimin PS, Matunin AP, Sakhno VG, Kimura K. Late Cretaceous to early Paleogene paleomagnetic results from Sikhote Alin, far eastern Russia: implications for deformation of East Asia. Earth and Planetary Science Letters. 1995; 130: 95–108.

[35] Otofuji Y, Matsuda T, Enami R, Uno K, Nishihama K, Halim N, Su L, Zaman H, Kulinich RG, Zimin PS, Matunin AP, Sakhno VG. Late Cretaceous palaeomagnetic results from Sikhote Alin, far eastern Russia: tectonic implications for the eastern margin of the Mongolia Block. Geophysical Journal International. 2003; 152: 202–214.

[36] Otofuji Y, Miura D, Takaba K, Takemoto K, Narumoto K, Zaman H, Inokuchi H, Kulinich RG, Zimin PS, Sakhno VG. Counter-clockwise rotation of the eastern part of the Mongolia block: Early Cretaceous palaeomagnetic results from Bikin, Far Eastern Russia. Geophysical Journal International. 2006; 164: 15–24.

Chapter 3

Stratigraphic and Petrological Insights into the Late Jurassic–Early Cretaceous Tectonic Framework of the Northwest Pacific Margin

Reishi Takashima, Hiroshi Nishi and
Takeyoshi Yoshida

Additional information is available at the end of the chapter

http://dx.doi.org/10.5772/intechopen.68289

Abstract

Late Jurassic–Early Cretaceous volcano-sedimentary sequences in the Sorachi, Kumaneshiri, and Yezo groups are exposed in central Hokkaido. The sequences are considered to reflect the Late Mesozoic tectonic history of the northwest Pacific continental margin. Based on the stratigraphic and petrological characteristics of igneous and volcaniclastic rocks of the Sorachi, Yezo, and Kabato groups, Late Jurassic–Early Cretaceous tectonics in central Hokkaido can be divided into six stages. Stage I (Tithonian) is characterized by extensive eruption of tholeiitic basalt accompanied with andesitic volcaniclastic rocks and terrigenous deposits. Seafloor spreading or large igneous province formation occurred near an island arc and/or continent during this stage. In Stage II, island arc volcanic islands were constructed on the basaltic rocks formed during Stage I. Stage III (latest Berriasian-Valanginian) is characterized by the formation of pull-apart basins accompanied by seafloor spreading. Widespread upwelling of the asthenosphere below central Hokkaido may have occurred during this stage. After the cessation of in situ volcanism in Stage IV (Hauterivian), submarine island arc volcanism reoccurred in Stage V (Barremian). In Stage VI (Aptian–Campanian), typical active continental margin volcanism occurred and voluminous granitic batholiths were formed in western Hokkaido.

Keywords: Jurassic, Cretaceous, Sorachi group, Kumaneshiri group, Asian continental margin, Hokkaido

1. Introduction

The Early Late Jurassic is characterized by a peak in submarine igneous activity represented by a negative peak in $^{87}Sr/^{86}Sr$ in marine sediments [7, 15]. The spreading rate of the mid-oceanic ridges increased in the Central Atlantic and Piemonte-Ligurian oceans [13, 37], when large igneous provinces (LIPs), such as the Shatsky Rise, formed in the Pacific. The active volcanism during this period was likely associated with significant global warming and transgression. Consequently, the Late Jurassic–Early Cretaceous was characterized by significant reconfiguration of the continents in East Asia. During this period, the Siberia Block (SB) collided with both the combined block of the Central Asian Orogenic Belt (CAOB)–North China Block (NCB)–South China Block (SCB) and the Omolon–Kolyma Block (OKB) along its southern and eastern margins, respectively [22, 36, 41]. These collisions resulted in the formation of the very large, amalgamated Asian continent and may have been reflected in the tendency for long-term increases in $^{87}Sr/^{86}Sr$ in marine sediments. The extensive upwelling of the asthenospheric mantle that occurred along the eastern Asian margin slightly after these collisions induced the renewal of the Asian continental crust, development of the NE–SW-trending extensional basins, and the extensive felsic volcanism referred to as the Late Yanshanian Event, which is considered to have started in the Late Jurassic and culminated in the late Early Cretaceous [14, 40].

The Japanese islands were part of the active continental margin of Asia until the opening of the Japan Sea in the Miocene. Of the Japanese islands, Hokkaido Island and the northeastern part of Honshu Island are considered to have been located on the supra-subduction magmatic zone–convergent boundary between the Paleo-Pacific Plate and the South China Block during the Early Cretaceous [27, 31]. The western part of Hokkaido Island is well suited for reconstructing the spatiotemporal characteristics of geodynamic evolution across the supra-subduction zone because the strata from the accretionary complex to the magmatic zones are intermittently well exposed. This paper examines the geodynamic processes of Late Jurassic–Early Cretaceous Hokkaido based on the sedimentary facies and geochemistry of volcaniclastic and igneous rocks.

2. Regional geologic setting

The Mesozoic rocks on western Hokkaido are divided into the N–S-trending tectonic divisions of the Oshima, Rebun–Kabato, and Sorachi–Yezo belts (**Figure 1**), which are considered to have formed along the west-dipping subduction zone of the Pacific oceanic plates under the eastern margin of the Asian continent [21].

The Oshima Belt consists mainly of a Jurassic accretionary complex and middle–Late Cretaceous granitic rocks (**Figure 2**). The Early Cretaceous volcano-sedimentary sequences, referred to as the Rebun and Kabato groups, are exposed along the eastern margin of the Oshima Belt. The distribution of the Rebun and Kabato groups is distinguished as the Rebun–Kabato Belt. The Sorachi–Yezo Belt is represented by a coherent succession consisting of the uppermost Jurassic ophiolite (Horokanai ophiolite), uppermost Jurassic–Lower Cretaceous submarine volcano-sedimentary sequences (Sorachi Group), and Lower Cretaceous–Paleogene submarine terrigenous sequences (Yezo Group), in ascending order (**Figure 2**). The early Cretaceous accretionary complexes are exposed in the central (Kamuikotan Zone) and eastern (Idonnappu Zone) parts of the Sorachi–Yezo Belt. Although both are accretionary

http://dx.doi.org/10.5772/intechopen.68289

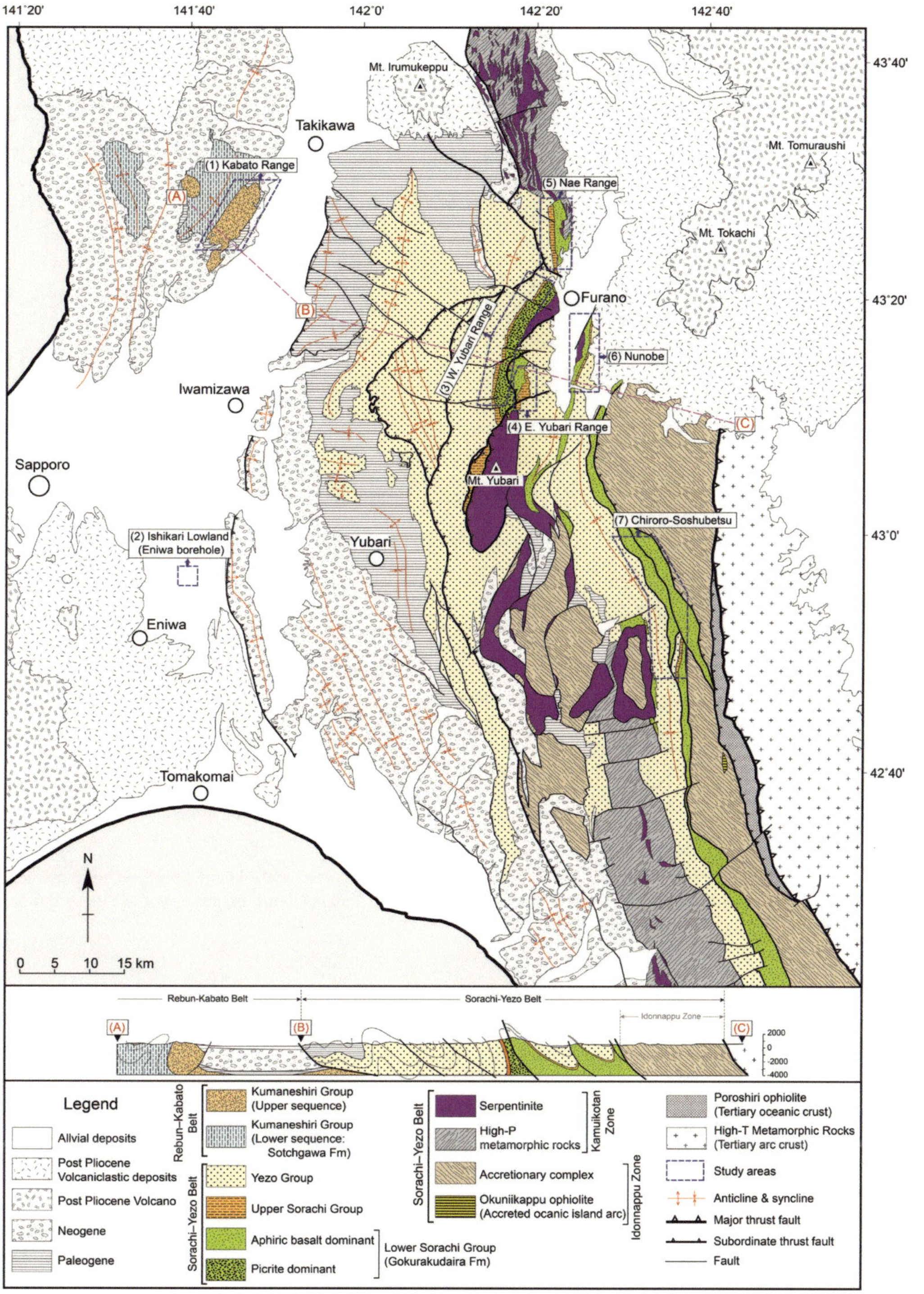

Figure 1. Geologic map and structural profile of central Hokkaido.

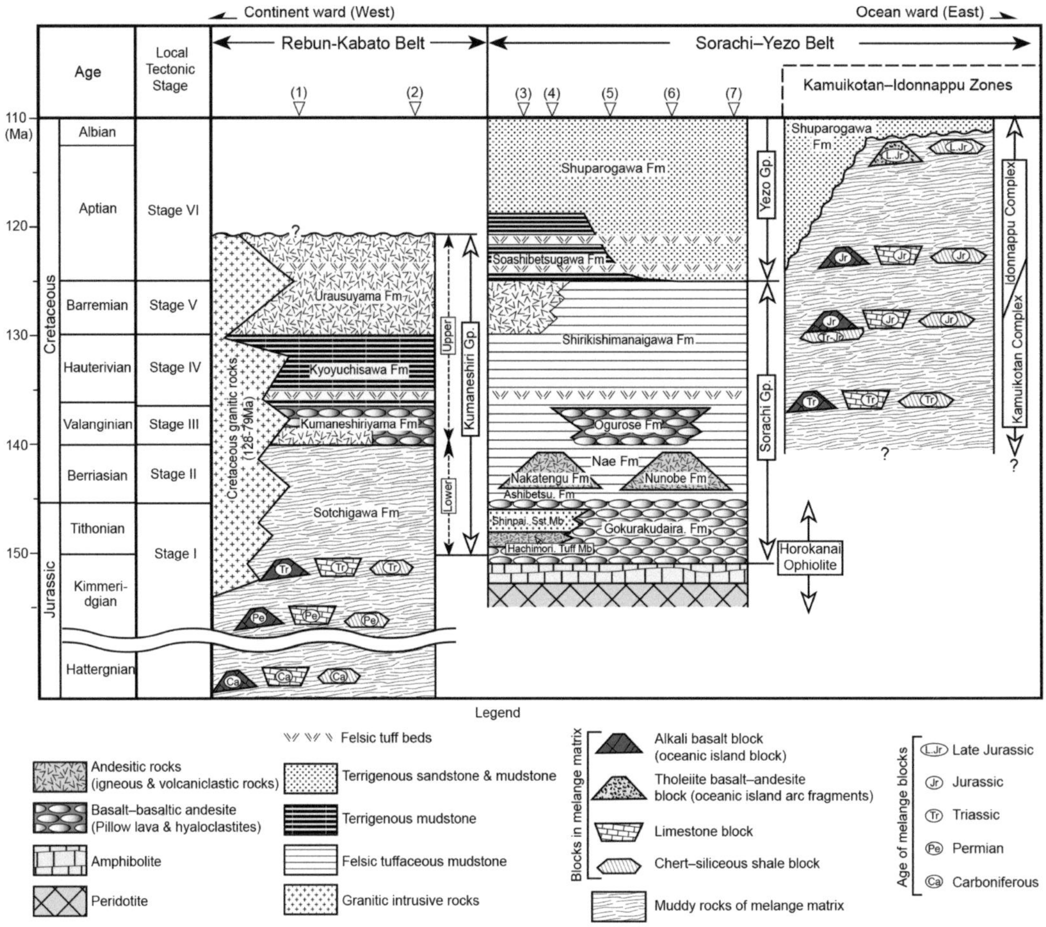

Figure 2. Schematic profile of Late Jurassic–Early Cretaceous arc-trench system in Hokkaido. Studied areas include (1) Kabato Range, (2) Ishikari Lowland (Eniwa borehole), (3) areas west of Yubari Range, (4) areas east of Yubari Range, (5) Nae Range, (6) Nunobe, and (7) Soshubetsu–Chiroro.

complexes showing a trend of becoming younger toward the east [11, 35], the former is high-pressure metamorphosed.

3. Stratigraphy of the upper Jurassic–Lower Cretaceous volcano-sedimentary succession in Hokkaido

Upper Jurassic–Lower Cretaceous volcano-sedimentary sequences of the Rebun–Kabato and Sorachi–Yezo belts occur in seven areas (**Figures 1–3**): (1) Kabato Range, (2) Ishikari Lowland (Eniwa borehole), (3) west of Yubari Range, (4) east of Yubari Range, (5) Nae Range, (6) Nunobe, and (7) Soshubetsu–Chiroro from west to east (**Figures 1** and **2**). The volcano-sedimentary sequences in the former two areas are named the Kumaneshiri Group, while the others are named the Sorachi and Yezo groups (**Figures 2** and **3**).

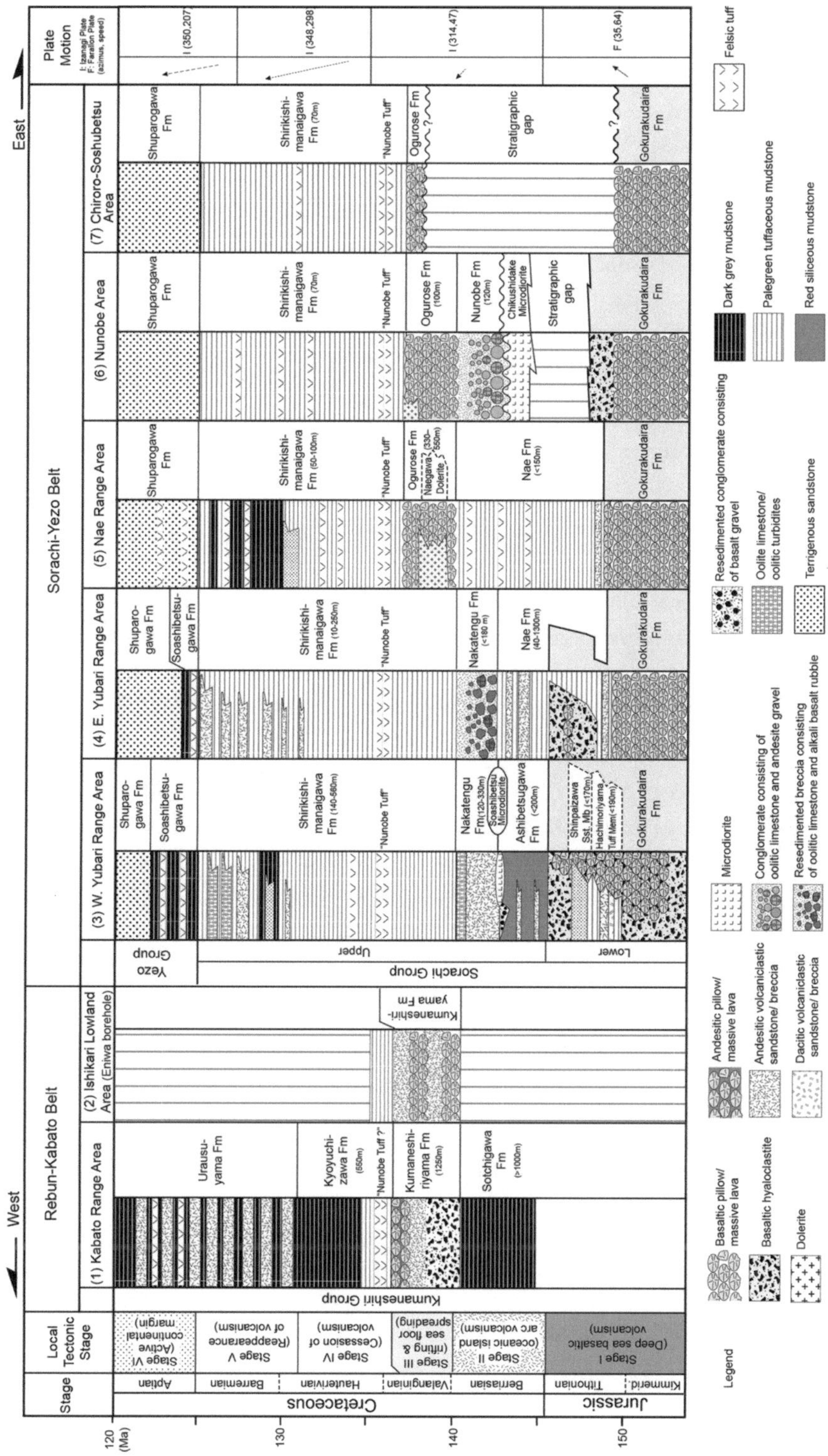

Figure 3. Generalized isochronous stratigraphic nomenclature of the Sorachi, Yezo, and Kumaneshiri groups in the seven studied sections (modified from Ref. [31]).

3.1. Kumaneshiri Group

The Kumaneshiri Group is represented by the Lower Cretaceous volcano-sedimentary sequences exposed in the Rebun–Kabato Belt, and consists mainly of submarine volcanic and sedimentary rocks (**Figure 3**). The group is made up of the Sotchigawa Formation, Kumaneshiriyama Formation, Kyoyuchizawa Formation, and Urausuyama Formation, in ascending order, and accompanied by basaltic-to-andesitic intrusive rocks.

The Sotchigawa Formation consists mainly of dark gray shale with occasional intercalations of sandstones and basaltic to rhyolitic tuff beds. Folding and faulting prevailed in these strata, which had a total thickness of approximately 2000 m. Radiolarian fossils assigning the Late Jurassic to Early Cretaceous period occur in this formation [12].

The Kumaneshiriyama Formation is mainly composed of volcaniclastic rocks with subordinate amounts of mudstone and pillow lava. The lava and rubble of the volcaniclastic rocks are olivine–clinopyroxene-phyric basalt and clinopyroxene–plagioclase-phyric basaltic andesite. Based on radiolarian biostratigraphy, the 500–1250 m-thick formation is correlative with the Late Berriasian to Valanginian [17, 26].

The Kyoyuchizawa Formation is a 600-m thick sequence consisting of dark gray and greenish gray mudstone with intercalations of rhyolitic fine-grained tuff beds. A 30-m sequence consisting of greenish gray mudstone frequently intercalated with rhyolitic tuff beds occurs at the basal part of this formation. This lithology resembles the "Nunobe Tuff Member" of the Sorachi Group described in Section 3.2.

The Urausuyama Formation is composed of alternating beds of basaltic-to-andesitic volcaniclastic sandstone and black mudstone. The thickness of this formation has been estimated to be approximately 1300–1600 m. Basaltic-to-andesitic volcanic breccia occurs intermittently in the upper part of this formation, and the rubble of volcanic breccia is composed of clinopyroxene–plagioclase-phyric basaltic andesite and clinopyroxene–plagioclase-phyric basaltic andesite. Radiolarian fossils obtained from the Kyoyuchizawa and Urausuyama formations imply that these formations are younger than Berriasian [17].

3.2. Sorachi Group

The Sorachi Group ranges from the Latest Jurassic to Barremian, and consists mainly of submarine volcanic and volcaniclastic rocks. Since the rocks of the Sorachi Group are laterally discontinuous [31], the Sorachi Group is divided into four rock assemblages based on the age and characteristics of the igneous rocks: (1) Uppermost Jurassic (tholeiitic basaltic rock assemblage), (2) Berriasian (andesitic rock assemblage), (3) Valanginian (basaltic rock assemblage), and (4) Hauterivian-Barremian (tuffaceous rock assemblage).

3.2.1. Uppermost Jurassic tholeiitic basaltic rock assemblage

The Gokurakudaira Formation and the lower part of the Nae Formation constitute this rock assemblage (Stage I rocks in **Figure 3**). The Gokurakudaira Formation is composed of basaltic pillow lava and hyaloclastite. Petrographic characteristics of the basalt in this formation differ among areas. For example, areas to the east of the Yubari Range, Nae Range, and Soshubetsu–Chiroro [areas (4), (5), and (7) in **Figures 1–3**] are composed of aphyric basalt, while those to the west of the Yubari Range [area (3) in **Figures 1–3**] are predominated by picrite. The basalt in the Nunobe area [area (6) in **Figures 1–3**] is mostly aphyric basalt, but the hyaloclastites in the uppermost part are olivine-phyric andesite. To the west of the Yubari Range, the Gokurakudaira Formation intercalates the Hachimoriyama Tuff Member and the Shinpaizawa Sandstone Member. The former member consists of tuffaceous mudstone and intercalates abundant dacitic-to-andesitic volcaniclastic rocks, while the latter is predominantly sandstone with alternating beds of terrigenous turbidite sandstone and hemipelagic mudstone. To the southeast of the Yubari Range, basaltic rocks intercalate thick tuffaceous mudstone that is correlated with the lower part of the Nae Formation. In this area, basalts below the tuffaceous mudstone consist mainly of pillow lava, while those above the tuffaceous mudstone are predominated by hyaloclastite.

The Nae Formation consists of pale green tuffaceous mudstone frequently intercalated with felsic tuff beds. However, the lower part of this formation rarely intercalates thin beds of andesitic volcaniclastic sandstone.

In terms of radiolarian biostratigraphy, the first occurrence of *Pseudodictyomitra carpatica,* which is assigned to be Tithonian, was identified at the lower part of the Hachimoriyama Tuff Member and the Nae Formation [31].

3.2.2. Berriasian andesitic rock assemblage

The Berriasian rock assemblage (Stage II rocks in **Figure 3**) of the Sorachi Group, which is constituted by the Ashibetsugawa, Nae, Nakatengu, and Nunobe formations, is accompanied with intrusive rocks of the Soashibetsu and Chikushidake microdiorites (**Figures 2** and **3**). Although [31] described the Soashibetsu and Chikushidake microdiorites as "micromonzonites," we have revised this here as the highly alkaline nature of these rocks may have resulted from secondary alteration.

To the west of the Yubari Range, the Ashibetsugawa Formation conformably overlies the basaltic hyaloclastite and/or pillow lava of the Gokurakudaira Formation. The Ashibetsugawa formation is composed mainly of red mudstone and frequently intercalates basaltic-to-andesitic volcaniclastic sandstone beds. The Nakatengu Formation, which covers conformably the Ashibetsugawa Formation, consists of andesitic volcaniclastic breccia. The basaltic conglomerate and oolitic turbidite beds occur at the base and top of this formation, respectively. The Soashibetsu microdiorite intrudes the border between the Ashibetsugawa and Nakatengu formations.

To the east of the Yubari and Nae ranges, the pale green tuffaceous mudstone of the Nae Formation conformably overlies the basaltic rocks of the Gokurakudaira Formation. Although the tuffaceous mudstone occurs throughout this interval in the Nae Range area, the Nakatengu Formation overlies the Nae Formation on the eastern side/to the east of the Yubari Range. The Nakatengu Formation on the east of the Yubari Range is composed of matrix-supported conglomerate beds containing sub-rounded rubble of alkali basalt and oolite limestones. Based on their sedimentological characteristics, the beds have been interpreted to be gravity flow deposits [31].

In the Nunobe area, Chikushidake microdiorite is unconformably covered by the basal conglomerate of the Nunobe Formation. The conglomerate is clast supported, and consists of subangular to sub-rounded rubble composed of andesite, oolite limestone, and microdiorite. The conglomerate grades into sandstone further upward [29].

3.2.3. Late Berriasian–Late Valanginian basaltic rock assemblage

Rocks of this interval are represented by the Ogurose Formation and the lower part of the Shirikishimanaigawa Formation (Stage III rocks of **Figure 3**). The Ogurose Formation occurs in the Nae Range and Nunobe areas, and consists of basaltic pillow lava and dolerite dikes. Pillow lobes are generally larger than those of the Gokurakudaira Formation and contain numerous bubbles. To the west and east of the Yubari Range, basaltic rocks are absent in the contemporaneous strata which consist mainly of pale green tuffaceous mudstone belonging to the lower part of the Shirikishimanaigawa Formation (**Figures 2** and **3**).

3.2.4. Late Valanginian–Barremian siliceous mudstone rock assemblage

The rocks of this interval are represented by the Shirikishimanaigawa Formation and are characterized by a predominance of pale green tuffaceous siliceous mudstone (Stage IV and V rocks of **Figure 3**). To the west of the Yubari Range, andesitic volcaniclastic sandstone beds and felsic tuff beds are intercalated in the upper part of the Shirikishimanaigawa Formation.

At an interval of several meters to several tens of meters, white siliceous felsic tuffs are intercalated frequently at the lower part of the Shirikishimanaigawa Formation. This interval is referred to here as the "Nunobe Tuff", and is traceable throughout the studied areas (**Figures 2** and **3**).

3.2.5. Post-Aptian rock assemblage

This rock assemblage is constituted by the Yezo Group, which ranges from the Aptian to the Paleocene [30] (Stage VI rocks of **Figure 3**). The Yezo Group consists mainly of hemipelagic mudstone and turbidite sandstone that were accumulated in a fore-arc basin along the active Asian continental margin. Since the Yezo Group yields abundant marine macro- and microfossils as well as felsic tuffs, numerous stratigraphic and paleoenvironmental studies have been carried out [1, 3, 8, 20, 24, 32–34].

The Yezo Group consists of the Soashibetsu, Shuparogawa, Maruyama, Hikagenosawa, Saku/Mikasa, Kashima/Haborogawa, and Hakobuchi formations, in ascending order [30]. The rhyolite fragments were taken from the volcaniclastic breccias which occur at the base of the Maruyama Formation. The Maruyama Formation is in the *Biticinella breggiensis* planktonic foraminiferal Zone, which ranges around 105–102 Ma [20, 24].

4. Geochemical analysis of igneous rocks

The igneous rocks and rubble of volcaniclastic rocks of the Kumaneshiri and Sorachi and Yezo groups were analyzed for major and trace elements. The samples from the Kumaneshiri Group in the Ishikari Lowland were obtained from the Eniwa SK-1 borehole, and were provided by Japan Petroleum Exploration Co., Ltd. (JAPEX).

Major and trace elements in the samples were identified using X-ray fluorescence spectrometry at the Faculty of Education, Fukushima University. FeO was analyzed by titration, and the H_2Ot contents were determined after ignition. Rare earth elements (REEs) were analyzed using an inductively coupled plasma mass spectrometer (ICP-MS) at the Geoanalytical Laboratory, Washington State University. Analytical results of the Gokurakudaira Formation were also obtained from Ref. [28]. All of the presented data are available from the Osaka Prefecture University Education and Research Archives (OPERA) (http://hdl.handle.net/10466/14732).

5. Results

Rocks of the Kumaneshiri and Sorachi groups consist of submarine volcanic and volcaniclastic rocks and mudstone; lithological correlation among areas was difficult because of lateral changes in lithology. Takashima et al. [31] correlated the sequences of the Sorachi Group with radiolarian data. Specifically, they examined the first occurrence of *P. carpatica* and the first and last occurrences of "*Cecrops*" *septemporatus*, and then divided the Sorachi Group into four stages based on lithologic features (Stages I–IV). Several horizons of the Kumaneshiri Group can also be correlated with these stages based on radiolarian biostratigraphy and the "Nunobe Tuff". Based on the lithological characteristics of the Kumanishiri Group, Stage IV of [31] can be subdivided into two stages because the interval of their stage IV includes the Kyoyuchizawa and Urausuyama formations in the Rebun–Kabato Belt. We redefine their Stage IV as being equivalent to the depositional interval of the Kyoyuchizawa Formation and the lower part of the Shirikishimanaigawa Formation (Late Valanginian–Hauterivian), and consider that Stage V can be correlated to the depositional interval of the Urausuyama Formation and the upper part of the Shirikishimanaigawa Formation (Barremian). Furthermore, we consider that Stage VI includes the deposition interval of the Yezo Group (Aptian–Paleocene).

Stratigraphic and areal changes in SiO_2–Nb/Y, a spider diagram of incompatible trace element abundance relative to N-MORB, and REE abundance relative to chondrite, are shown in **Figures 4–6**.

5.1. Rocks of Stage I (Kimmeridgian–Tithonian): Gokurakudaira Formation and the lower part of the Nae Formation

The basaltic rocks of the Gokurakudaira Formation consist mainly of tholeiitic basalts, which have a mid-ocean ridge basalt (MORB)-like flat pattern in N-MORB-normalized incompatible trace-element abundance (**Figures 4** and **5**). Conversely, the picrites and olivine-phyric andesites from the areas to the west of the Yubari Range and Nunobe are depleted in all elements, N-MORB-normalized incompatible trace elements, and REE abundances. Compared to heavy REEs (HREEs), the picrites and olivine-phyric andesites are also depleted in light REEs (LREEs) (**Figures 5** and **6**). Although basaltic rocks occur in two horizons to the east of the Yubari Range, their compositions are similar.

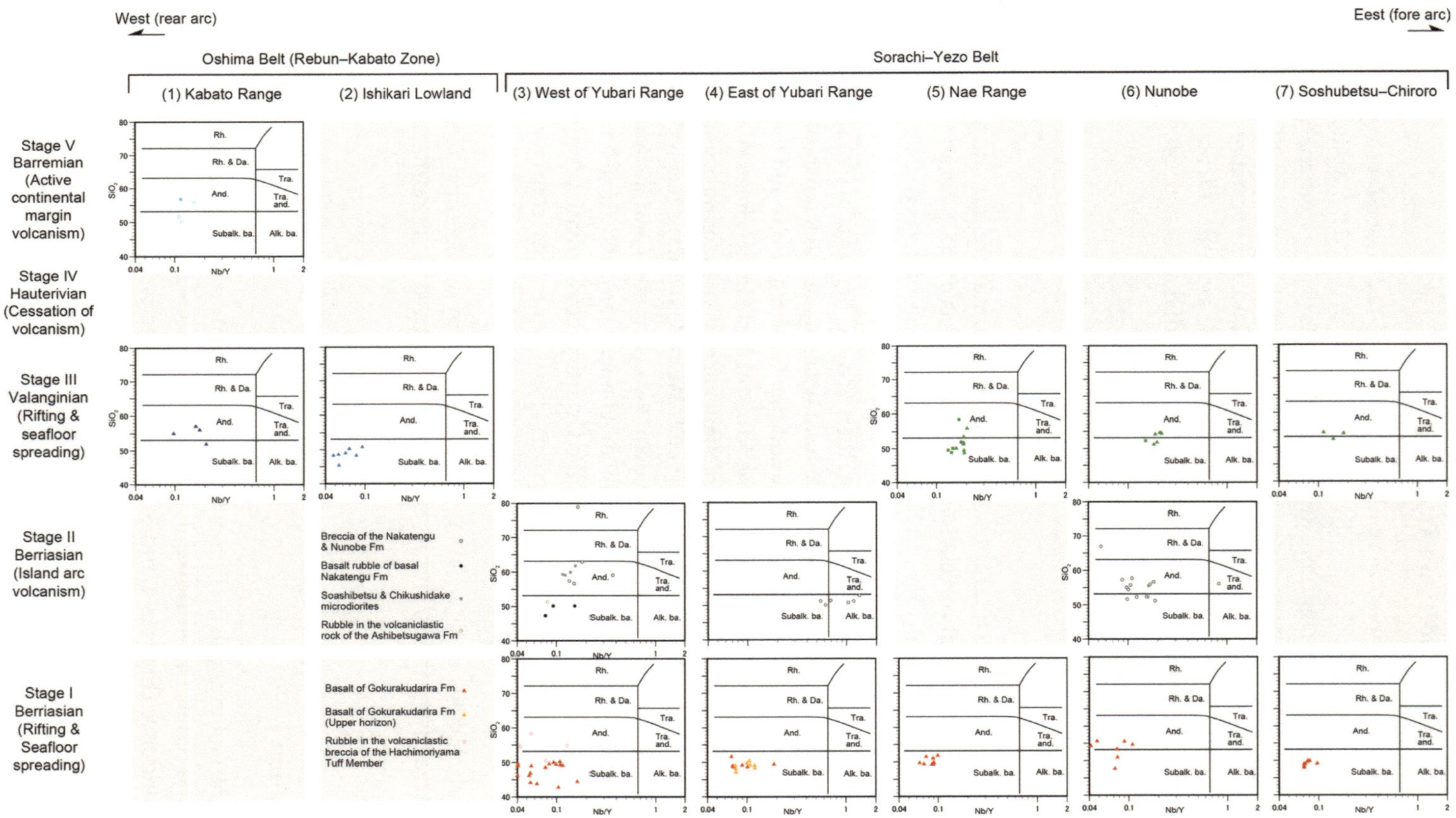

Figure 4. Plot of SiO_2 versus Nb/Y for igneous rocks from the Sorachi and Kumaneshiri groups in the studied sections. Fields are from Ref. [39]. Triangular, square, and circular symbols show extrusive rocks, intrusive rocks, and rubble in the volcaniclastic rocks, respectively.

http://dx.doi.org/10.5772/intechopen.68289

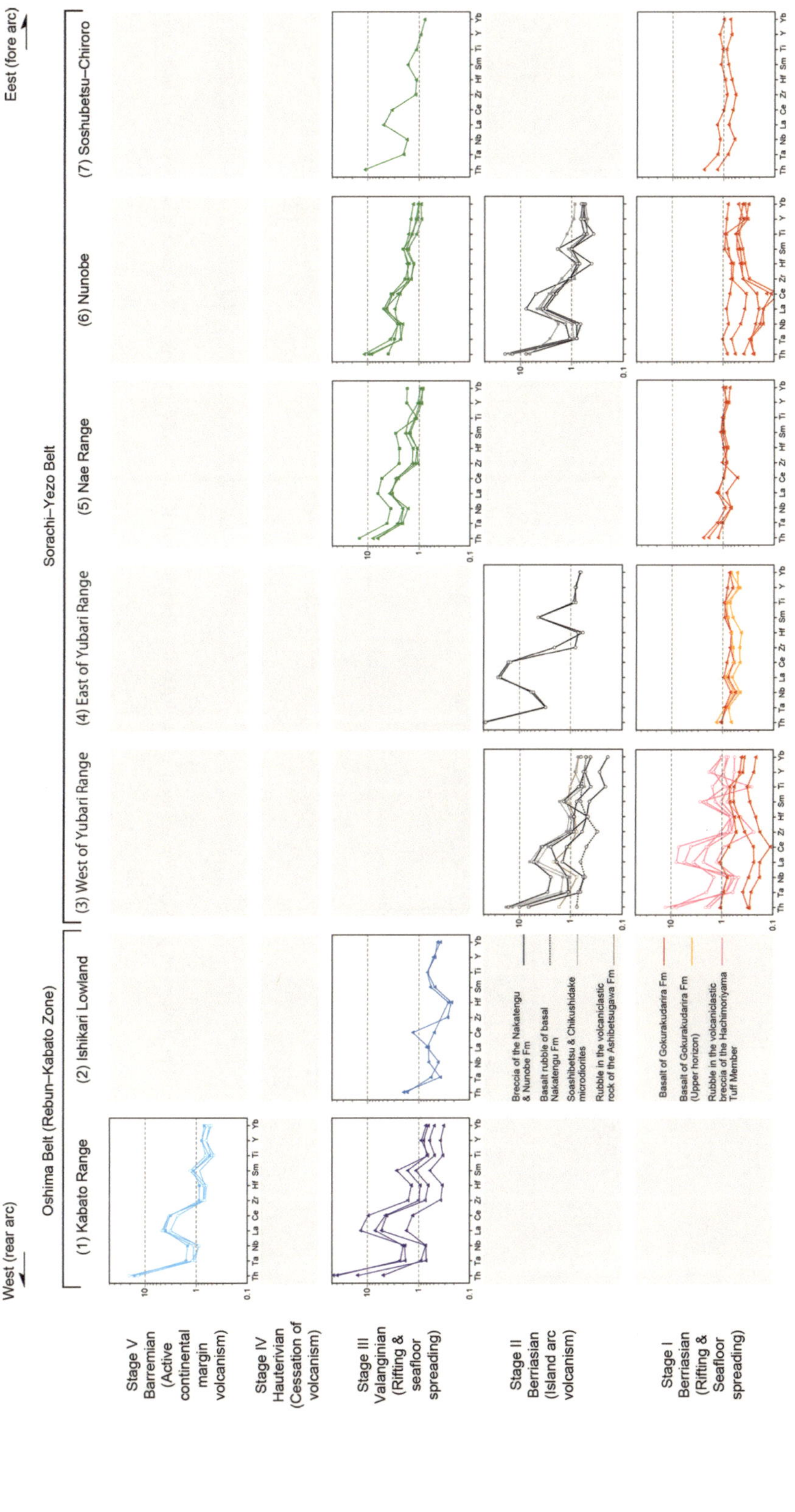

Figure 5. N-MORB-normalized incompatible trace element patterns [25] for igneous rocks from the Sorachi and Kumaneshiri groups in the studied sections. Symbols as in **Figure 4**.

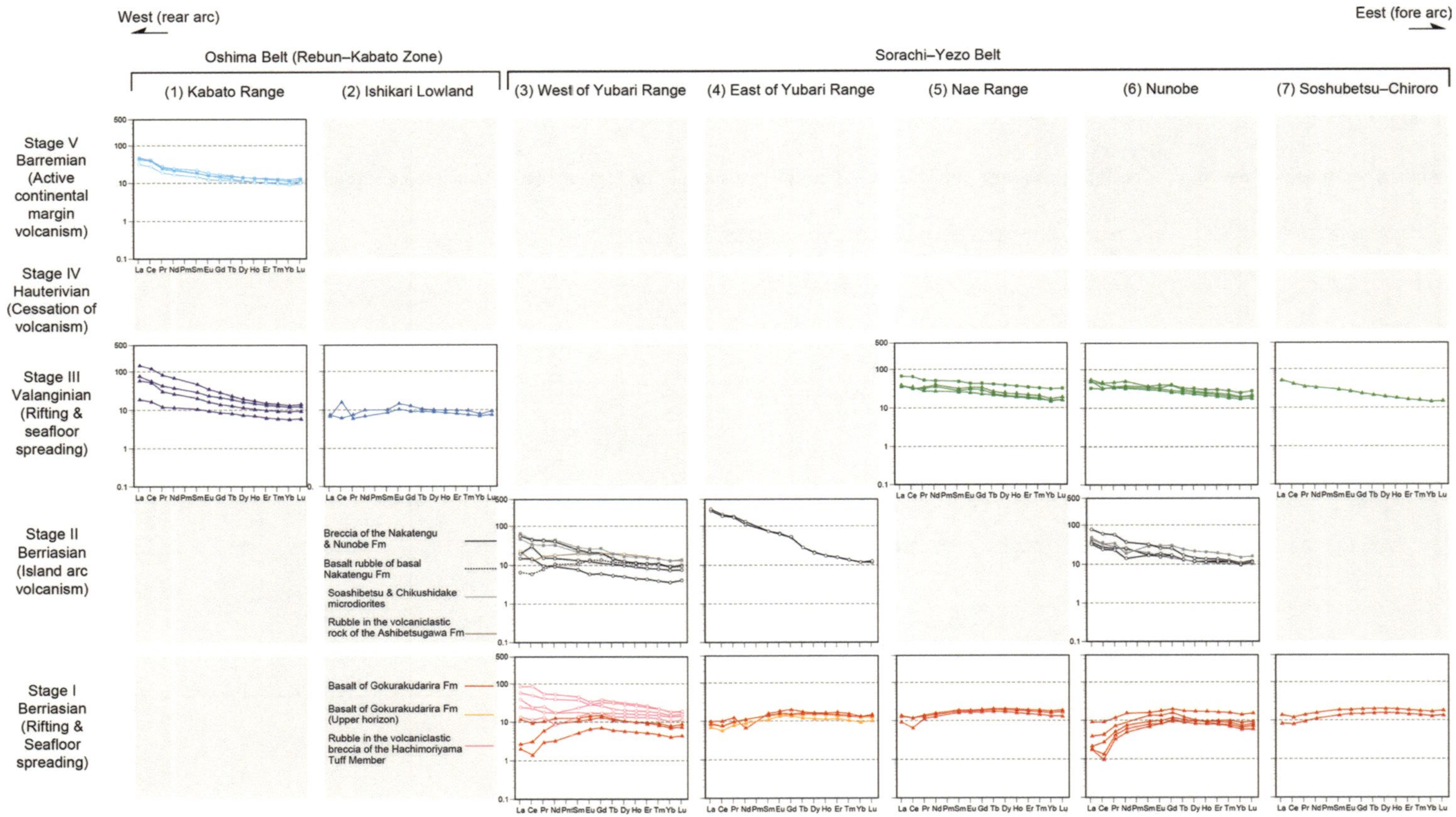

Figure 6. Chondrite-normalized rare earth element patterns for volcanic rocks from the Sorachi and Kumaneshiri groups in the studied sections. Symbols as in **Figure 4**.

The essential lenses in the volcaniclastic rocks in the Hachimoriyama Tuff Member occur mostly in calc-alkaline andesite (**Figures 4** and **8**). The N-MORB-normalized incompatible trace element abundance patterns are characterized by the enrichment of large-ion lithophile elements (LILEs) and depletion of high-field strength elements (HFSEs), with negative anomalies in Ta and Nb abundances relative to N-MORB (**Figure 5**). Chondrite-normalized REE abundance patterns show higher LREE content compared to the content of HREEs (**Figure 6**).

5.2. Rocks of Stage II (Berriasian)

Basalt rubble and gravels from the volcaniclastic rocks in the Ashibetsugawa Formation and basaltic conglomerate of the basal Nakatengu Formation are tholeiitic basalt, showing flat patterns in N-MORB-normalized incompatible trace elements and chondrite-normalized REE patterns resembling those of the basaltic rocks of the underlying Gokurakudaira Formation (**Figures 5** and **6**).

The rubble of volcaniclastic breccia and volcanogenic conglomerate of the Nakatengu and Nunobe formations, as well as the rocks of the Soashibetsu and Chikushidake microdiorites in the areas to the west of the Yubari Range and Nunobe, is plotted in the fields of calc-alkaline andesite (**Figures 4** and **8**). Conversely, the rubble of volcanogenic breccia of the Nakatengu Formation to the east of the Yubari Range is alkali basalt (**Figure 5**). The calc-alkaline andesites and alkali basalts are characterized as being rich in LILEs relative to HFSEs having Ta and Nb trough in the N-MORB-normalized incompatible trace element pattern (**Figure 6**). For the chondrite-normalized REE pattern, LREEs are considerably higher than HREEs. In particular, the alkali basalts of the Nakatengu Formation to the east of the Yubari Range exhibit high LREE/HREE ratios.

5.3. Rocks of Stage III (Late Berriacian-Late Valanginian)

Volcanic and intrusive rocks of the Ogurose Formation of the Sorachi Group and Kumaneshiriyama Formation of the Kumaneshiri Group are mostly occurred in fields of

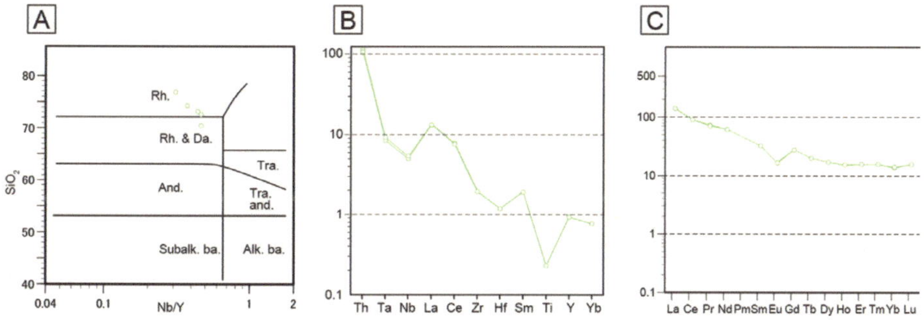

Figure 7. SiO_2 versus Nb/Y diagram (A) and N-MORB-normalized incompatible trace element (B), and chondrite-normalized rare earth element (C) patterns from the Yezo Group. Symbols as in **Figure 4**.

calc-alkaline basalt to andesite, while rocks of the Kumaneshiriyama Formation from the Eniwa borehole are classified as tholeiite basalt (**Figures 4** and **8**).

In the diagram of the N-MORB-normalized incompatible trace element abundances of this stage (**Figure 5**), the igneous rocks of the Sorachi Group and Kumaneshiriyama Formation in the Kabato Range exhibit a high concentration of LILEs relative to that of HFSEs, with a negative anomaly in Ta and Nb contents. **Figure 5** shows that the high LILE/HFSE ratio and depletion in Ta and Nb are prominent in the Kabato Range area, while the tholeiitic basalts of the Kumaneshiriyama Formation at the Eniwa borehole exhibit a flat pattern and are generally depleted in abundance (**Figure 5**). In the chondrite-normalized REE abundance diagram (**Figure 6**), the tholeiite basalts at the Eniwa borehole are the most depleted in this stage and exhibit a flat pattern, while the igneous rocks from other areas are more enriched in abundance and have slightly higher LREE/HREE ratios.

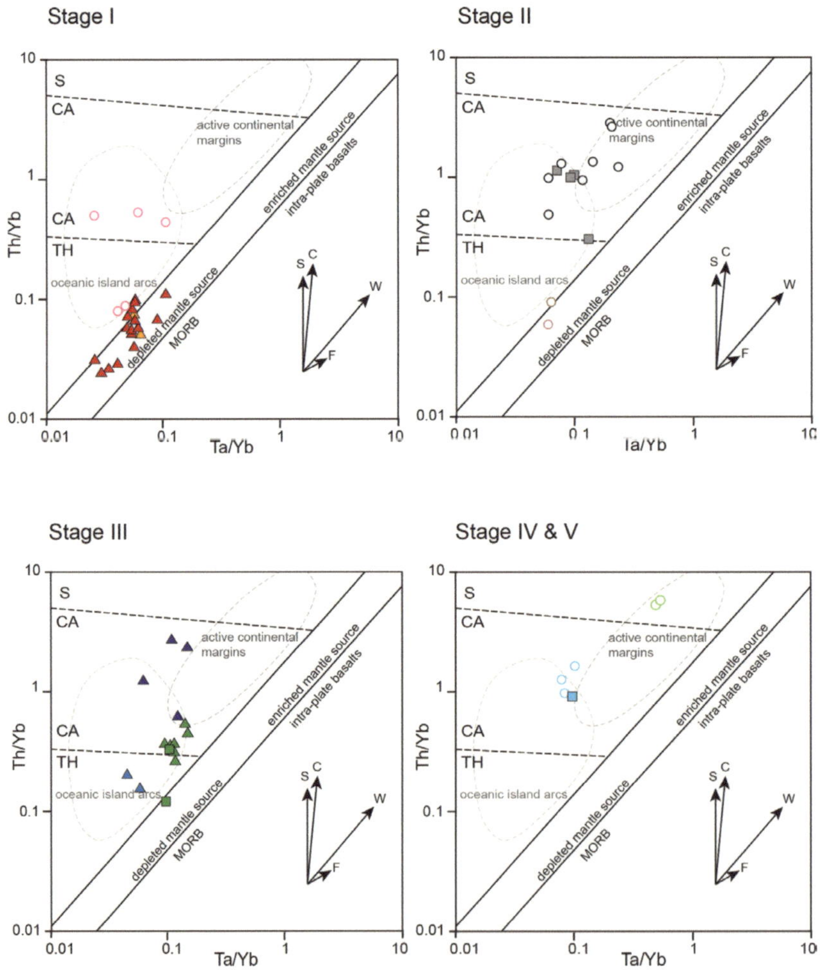

Figure 8. Plot of Th/Yb versus Ta/Yb for igneous rocks from the Sorachi, Kumaneshiri, and Yezo groups. Fields and arrows in the diagram are from Refs. [23, 38]. Symbols as in **Figure 4**.

5.4. Rocks of Stage V (Barremian)

The lava and volcanic breccia of the Urausuyama Formation in Kabato Range occur in fields of calc-alkaline basalt to andesite. The diagram of the N-MORB-normalized incompatible trace element patterns shows high LILE and low HFSE contents with marked depletion of Ta and Nb (**Figure 5**). The chondrite-normalized REE patterns show a high concentration of LREEs relative to that of HREEs (**Figure 6**). In the Ta/Tb-Tb/Yb diagram, rocks of this stage occur in the field between the oceanic island arc and active continental margin settings (**Figure 8**).

5.5. Rocks of Stage VI (Aptian-Campanian)

The rhyolite fragments from the Maruyama Formation of the Yezo Group have high LILE/HFSE and LREE/HREE ratios in N-MORB-normalized incompatible trace element and chondrite-normalized REE patterns (**Figure 7**). These rhyolites occur in the active continental margin setting in the Ta/Tb-Tb/Yb diagram (**Figure 8**).

6. Discussion

In this section, we discuss the tectonic models of western-central Hokkaido based on the stratigraphic and petrologic characteristics of the Sorachi and Kumaneshiri groups.

6.1. Stage I (Late Jurassic)

This stage was characterized by extensive eruption of N-MORB-like tholeiitic basalt represented by the Gokurakudaira Formation. The origin of these basaltic rocks is controversial, and several tectonic models have been proposed. Most of the models attributed the origin of these basalts to the mid-ocean ridge, back-arc basin, or oceanic plateau [4, 9, 16, 19, 28, 31].

Although most of the basaltic rocks of the Gokurakudaira Formation have similar geochemical characteristics to those of the N-MORB and back-arc basin, the petrological characteristics of the picrite on the western side of the Yubari Range show extremely high-temperature magmatism, similar to that of Archean komatiite [4]. On the other hand, the basaltic rocks of the Gokurakudaira Formation intercalate the Hachimoriyama Tuff Member, which contains andesitic to dacitic volcaniclastic rocks. The essential andesite lenses show typical supra-subduction zone affinity, with negative anomalies in both Ta and Nb in N-MORB-normalized incompatible trace element patterns (**Figure 5**). These findings suggest that the basaltic rocks of the Gokurakudaira Formation erupted near the island arc. Furthermore, terrigenous sandstone and mudstone also co-occur in this formation, suggesting that extensive eruption of MORB-like basalt and komatiite-like picrite occurred near the continental crust and/or island arc setting.

In order to explain such constraints, the three models shown in **Figure 9** are considered. The first model proposes that the Gokurakudaira Formation was formed by seafloor spreading with slab roll back and asthenospheric upwelling [31]. Takashima et al. [31] considered that

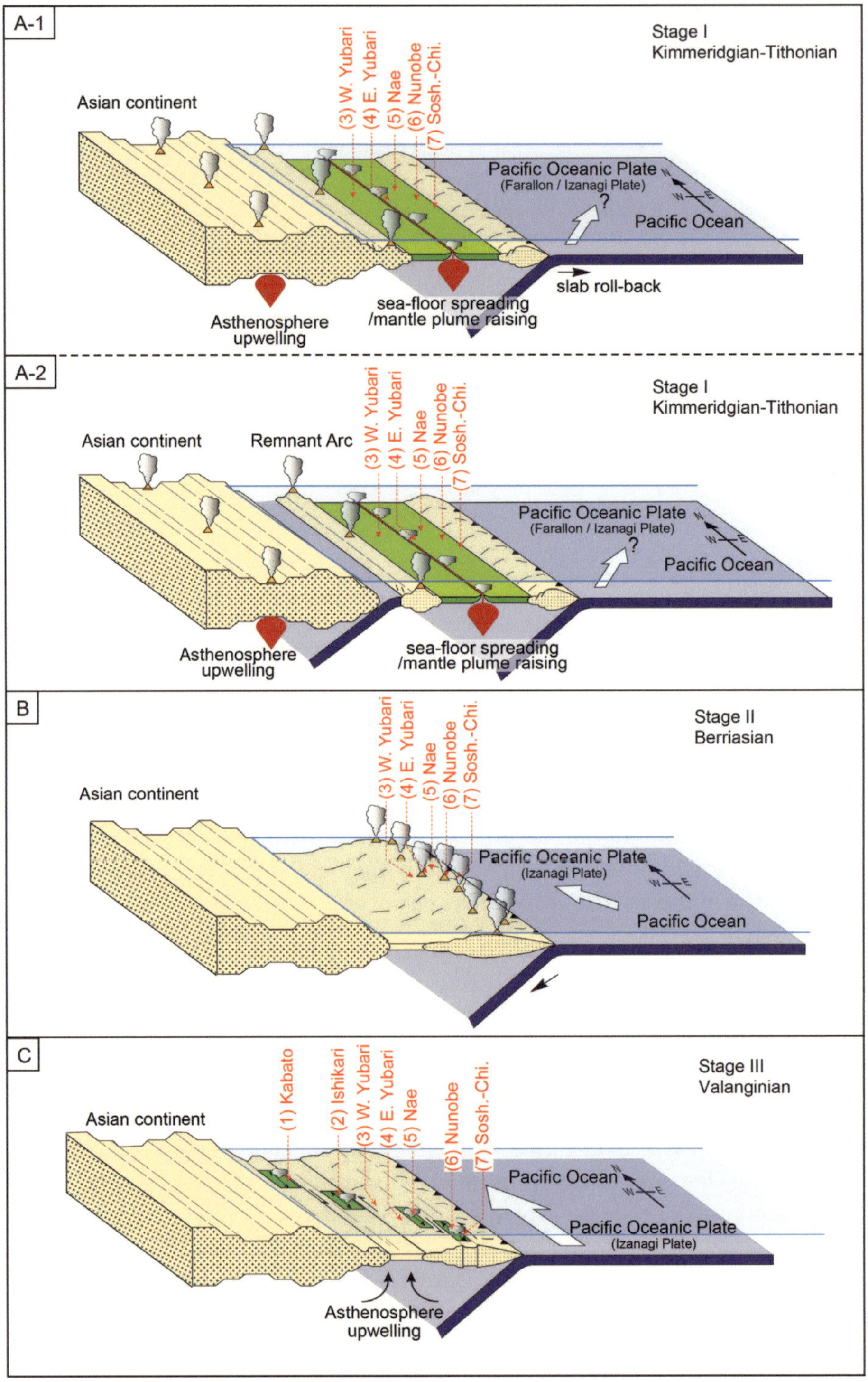

Figure 9. Tectonic models of Late Jurassic-Early Cretaceous Hokkaido. A-1 and A-2 are two plausible models of Stage I (Late Jurassic). B and C are tectonic models of Stage II (Berriasian) and Stage III (Valanginian).

this asthenospheric upwelling could be attributed to the Late Yanshianian Event. In southeast China, back-arc extension and injection of anomalously high-temperature (>1200°C) melts into the lower crust are considered to have occurred during the Late Jurassic [6]. As shown in **Figure 9A-2**, another model of the accretion of oceanic remnant-arc–back-arc basin can also be explainable (**Figure 9A-2**), although no collision zone or remnant arc has been found to date.

On the other hand, the oceanic plateau is also considered to be a strong candidate for the origin of the basaltic rocks of the Gokurakudaira Formation [4, 9, 16]. Large igneous province formation and interactions between depleted mantle and continental crust could also explain the coexistence of basalts of the Gokurakudaira Formation, as well as the andesitic volcaniclastic rocks of the Hachimoriyama Tuff Member and terrigenous sediments of the Shinpaizawa Sandstone Member (**Figure 9A-1**, **A-2**). In the case of the Kerguelen Plateau during the Cretaceous, some basaltic rocks have geochemical signatures that are consistent with a minor contribution from continental crustal material [2, 18]. These rocks also exhibit a prominent trough of Nb and Ta in primitive-mantle-normalized incompatible element abundance patterns, in common with the andesitic rocks of the Hachimoriyama Tuff Member.

6.2. Stage 2 (Berriasian)

The Berriasian Stage is characterized by the formation of andesitic to alkali basaltic volcanoes on the western and eastern sides of the Yubari Range and Nunobe areas. These volcanoes were partly emerged above the sea surface because gravels of oolitic limestones as well as oolitic turbidite beds, which are typically formed in shallow warm intertidal environments, are contained in the volcaniclastic sediments [31].

The volcanoes were formed on the supra-subduction zone because the andesite and microdiorites of this stage show typical island arc characteristics in spider diagrams of N-MORB-normalized incompatible element patterns (**Figure 5**). In the diagram showing Th/Yb-Ta/Yb ratios (**Figure 8**), igneous rocks of this stage were influenced by fluids in the subducting slab and clearly occur in the field of oceanic island arc areas.

Numerous intrusive bodies of microdiorite and diorite occur in the serpentinite bodies of the Horokanai Ophiolite throughout the axial zone of Hokkaido [5]. Although there is no age constraint for these intrusive rocks, N–S trending of microdiorite bodies might indicate that N–S-trending chains of andesitic volcanic edifices were formed in the Sorachi–Yezo Belt during this period (**Figure 9B**).

6.3. Stage III (Late Berriacian–Late Valanginian)

In the Sorachi-Yezo Belt, the emerged andesitic volcanoes subsided into deep marine environments and basaltic-to-basaltic andesitic pillow lava erupted in the areas of the Nae Range, Nunobe, and Soshubetsu–Chiroro. Contemporaneously, submarine eruption of basalt and andesite occurred in the Rebun–Kabato Zone which is predominated by the Kumaneshiriyama Formation.

Basalts and basaltic andesites in the Sorachi–Yezo Belt show typical island arc affinity in N-MORB-normalized incompatible trace element abundance patterns, which exhibit LILE enrichment relative to HFSEs and Nb- and Ta-negative anomalies (**Figure 5**). However, the influence of subduction components is less apparent than those of the previous Stage 2 rocks, as shown in **Figure 8**. The N-MORB-normalized incompatible trace element and chondrite-normalized REE patterns of this stage show that basalts of the Ishikari Lowland are most depleted in both diagrams (**Figures 5** and **6**). Takashima et al. [31] proposed that the formation of pull-apart basins accompanied by seafloor spreading occurred in the Sorachi-Yezo Belt. In conjunction with the present geochemical results, their findings show that asthenospheric upwelling and crustal thinning were most prominent in the Ishikari Lowland, and that the effect of island arc and/or continental crust material increased toward the east and west (**Figure 9C**).

Contemporaneous basaltic-to-rhyolitic volcanism has been recorded along the eastern margin of the Tohoku region (e.g., Yamadori and Harachiyama Formations), suggesting that widespread asthenospheric upwelling occurred along the eastern margin of the continent during this period. Paleomagnetic studies of the Sorachi Group have shown that rocks of Stages I–III were located several thousands of kilometers south of the present position [10].

6.4. Stage IV (Late Valanginian–Hauterivian)

In this period, in situ volcanic activity decreased in both the Rebun–Kabato Zone and the Sorachi–Yezo Belt. Takashima et al. [31] considered that the cessation of volcanism during this period was caused by strike slip movement in the subduction zone.

6.5. Stage V (Late Hauterivian–Barremian)

This stage is characterized by the reappearance of volcanism in the Rebun–Kabato Zone. Although there is no evidence of in situ volcanism in the Sorachi–Yezo Belt, the volcaniclastic turbidities occur frequently on the eastern and western sides of the Yubari Range (**Figure 3**). In the diagram of Th/Yb-Ta/Yb ratios (**Figure 8**), the volcanic rocks of this stage are characterized as being transitional between the oceanic island arc and active continental margins.

6.6. Stage VI (Aptian–Campanian)

The Yezo Group, a fore-arc basin sequence consisting of terrigenous turbidite sandstone and hemipelagic mudstones, started to accumulate in the Sorachi–Yezo Belt in this stage. Abundant felsic tuff beds are also intercalated in the Yezo Group, and the rhyolitic and granitic rocks that formed during this period exhibit a typical continental arc setting (**Figure 8**). In this stage, numerous granitic batholiths were formed in the Oshima Belt as well as along the eastern margin of the Tohoku region.

Acknowledgements

We sincerely thank Jun-ichi Kimura and Yoshitaka Nagahashi for support with the X-ray fluorescence analyses at the Fukushima University, and to Japan Petroleum Exploration Co., Ltd.

for providing samples of the Eniwa SK-1 borehole core. This work was supported by JSPS KAKENHI (Grant Nos. JP24244082 and JP25287130).

Author details

Reishi Takashima[1]*, Hiroshi Nishi[1] and Takeyoshi Yoshida[1,2]

*Address all correspondence to: rtaka@m.tohoku.ac.jp

1 The Center for Academic Resources and Archives, Tohoku University Museum, Tohoku University, Sendai, Japan

2 Institute of Earth Sciences, Graduate School of Science, Tohoku University, Sendai, Japan

References

[1] Du Vivier ADC, Selby D, Condon DJ, Takashima R, Nishi H. Pacific 187Os/188Os isotope chemistry and U–Pb geochronology: Synchroneity of global Os isotope change across OAE 2. Earth Planetary Science Letters. 2015;**428**:204–216

[2] Frey FA, Weis D, Borisova AY, Xu G. Involvement of continental crust in the formation of the Cretaceous Kerguelen Plateau: New perspectives from ODP Leg 120 sites. Journal of Petrology. 2002;**43**:1207–1239

[3] Hasegawa T. Cenomanian–Turonian carbon isotope events recorded in terrestrial organic matter from northern Japan. Palaeogeography, Palaeoclimatology, Palaeoecology. 1997;**130**:251–273.

[4] Ichiyama Y, Ishiwatari A, Kimura J, Senda R, Kawabata H, Tatsumi Y. Picrites in central Hokaido: Evidence of extremely high temperature magmatism in the Late Jurassic ocean recorded in an accreted oceanic plateau. Geology. 2012;**40**:411–414

[5] Igi S, Tanaka K, Hara M, Sato H. Geology of the Horokanai District, with Geological Sheet Map at 1:50,000. Kawasaki: Geological Survey of Japan. 1958: 55 p.

[6] Jiang Y-H, Ling H-F, Jiang S-Y, Fan H-H, Shen W-Z, Ni P. Petrogenesis of a Late Jurassic peraluminous volcanic complex and its high-Mg, potassic, quenched enclaves at Xiangshan, southeast China. Journal of Petrology. 2005;**46**:1121–1154

[7] Jones CE, Jenkyns HC. Seawater strontium isotopes, oceanic anoxic events and seafloor hydrothermal activity in the Jurassic and Cretaceous. American Journal of Science. 2001;**301**:112–149

[8] Kawabe F. Cretaceous stratigraphy in the Oyubari area, central Hokkaido, Japan. Bulletin of the National Science Museum. 2000;**26**(Series C, Geology):9–56

[9] Kimura G, Sakakibara M, Okamura M. Plumes in central Panthalassa? Deductions from accreted oceanic fragments in Japan. Tectonics. 1994;**13**:905–916

[10] Kitagawa Y, Takashima R, Itoh Y. Paleomagnetism of the Sorachi and Yezo Group in the Ashibetsu area, central Hokkaido, Japan. Bulletin of Tohoku University Museum. 2016;(15):109–125

[11] Kiyokawa S. Geology of the Idonnappu Belt, central Hokkaido, Japan: Evolution of a Cretaceous accretionary complex. Tectonics. 1992;**11**:1180–1206

[12] Kondo H. Stratigraphy and geological structure of the Kumaneshiri Group in the Kabato Mountains, Hokkaido, Japan. Journal of Geological Society of Japan. 1991;**97**:357–376

[13] Labails C, Olivet J-L, Aslanian D, Roest WR. An alternative early opening scenario for the Central Atlantic Ocean. Earth and Planetary Science Letters. 2010;**297**:355–368

[14] Li X-H, Chung S-L, Zhou H, Lo C-H, Liu Y, Chen C-H. Jurassic intraplate magmatism in southern Hunan-eastern Guangxi: 40Ar/39Ar dating, geochemistry, Sr-Nd isotopes and implications for the tectonic evolution of SE China. In: Malpas J, Fletcher CJN, Ali JR, Aitchison JC, editors. Aspect of the Tectonic Evolution of China. Geological Society. Vol. 226. London: Special Publications; 2004. pp. 193–215

[15] McArthur JM, Howarth RJ, Bailey TR. Strontium isotope stratigraphy: LOWESS version 3: Best fit to the marine Sr-isotope curve for 0–509 Ma and accompanying look-up table for deriving numerical age. The Journal of Geology. 2001;**109**:155–170

[16] Nagahashi T, Miyashita S. Petrology of the greenstones of the lower Sorachi Group in the Sorachi-Yezo belt, central Hokkaido, Japan, with special reference to discrimination between oceanic plateau basalts and mid-oceanic ridge basalts. The Island Arc. 2002;**11**:122–141

[17] Nagata M, Kito N, Niida K. The Kumaneshiri Group in the Kabato Mountains: The age and nature as an Early Cretaceous volcanic arc. In: Editorial Committee of Geology and Tectonics of Hokkaido, editors. Geology and Tectonics of Hokkaido. Sapporo: Monograph of the Association for the Geological Collaboration in Japan. Vol. 31. 1986. pp. 379–402

[18] Neal CR, Mahoney JJ, Chazey WJ. Mantle sources and the highly variable role of continental lithosphere in basalt petrogenesis of the Kerguelen Plateau and Broken Ridge LIP: Results from ODP Leg 183. Journal of Petrology. 2002;**43**:1177–1205

[19] Niida K, Kito N. Cretaceous arc–trench systems in Hokkaido. In: Editorial Committee of Geology and Tectonics of Hokkaido, editors. Geology and Tectonics of Hokkaido. Sapporo: Monograph of the Association for the Geological Collaboration in Japan. Vol. 31. 1986. pp. 379-402.

[20] Nishi H, Takashima R, Hatsugai T, Saito T, Moriya K, Ennyu A, Sakai T. Planktonic foraminiferal zonation in the Cretaceous Yezo Group, Central Hokkaido, Japan. Journal of Asian Earth Sciences. 2003;**21**:867–886

[21] Okada H. Migration of ancient arc-trench systems. In: Dott RJH, Shaver RH, editors. Modern and Ancient Geosynclinals Sedimentation. Society of Economic Paleontologists and Mineralogists, Special Publication No. 19; 1974. pp. 311–320

[22] Oxman VS. Tectonic evolution of the Mesozoic Verkhoyansk-Kolyma belt (NE Asia). Tectonophysics. 2003;**365**:45–76

[23] Pearce JA. Trace element characteristics of lavas from destructive plate boundaries. In: Thorpe ES, editors. Andesites. New York, NY: John Wiley and Sons; 1982. pp. 525–548

[24] Quidelleur X, Paquette JL, Fiet N, Takashima R, Tiepolo M, Desmares D, Nishi H, Grosheny D. New U-Pb (ID-TIMS and LA-ICPMS) and 40Ar/39Ar geochronological constraints of the Cretaceous geologic time scale calibration from Hokkaido (Japan). Chemical Geology. 2011;**286**:72–83

[25] Sun S-S, McDounough WF. Chemical and isotopic systematics of oceanic basalts: implications for mantle composition and processes. In: Saunders AD, Norry MJ, editors. Magmatism in the Ocean Basins. Geological Society, London, Special Publications. Vol. 42; 1989. pp. 313–345

[26] Suzuki N, Kurita H, Takashima R. Earliest Cretaceous radiolarian assemblages from a deep borehole section in the southern Ishikari Plain, central Hokkaido, and their implications. In: The 106th Annual Meeting of the Geological Society of Japan; October 9-10; Nagoya. Tokyo: The Geological Society of Japan; 1999. p. 307.

[27] Taira A. Tectonic evolution of the Japanese island arc system. Annual Review of Earth and Planetary Sciences. 2001;**29**:109–134

[28] Takashima R, Nishi H, Yoshida T. Geology, petrology and tectonic setting of the Late Jurassic ophiolite in Hokkaido, Japan. Journal of Asian Earth Sciences. 2002;**21**:197–215

[29] Takashima R, Nishi H, Miyamoto Y, Yoshida T. Geology and stratigraphy of the Sorachi and Yezo groups in the Tokyo University Forests in Hokkaido, Japan. Bulletin of Tokyo University Forests. 2002;**108**:57–76

[30] Takashima R, Kawabe F, Nishi H, Moriya K, Wani R, Ando H. Geology and stratigraphy of forearc basin sediments in Hokkaido, Japan: Cretaceous environmental events on the north-west Pacific margin. Cretaceous Research. 2004;**25**:365–390

[31] Takashima R, Nishi H, Yoshida T. Late Jurassic–Early Cretaceous intra-arc sedimentation and volcanism linked to plate motion change in northern Japan. Geological Magazine. 2006;**143**:753–770

[32] Takashima R, Nishi H, Yamanaka T, Hayashi K, Waseda A, Obuse A, Tomosugi T, Deguchi N, Mochizuki S. High-resolution terrestrial carbon isotope and planktonic foraminiferal records of the Upper Cenomanian to the Lower Campanian in the Northwest Pacific. Earth and Planetary Science Letters. 2010;**289**:570–582

[33] Takashima R, Nishi H, Yamanaka T, Tomosugi T, Fernando AG, Tanabe K, Moriya K, Kawabe F, Hayashi K. Prevailing toxic environments in the Pacific Ocean during the mid-Cretaceous Oceanic Anoxic Event 2. Nature Communications. 2011;**2-234.** DOI: 10.1038/ncomms1233

[34] Taketani, Y. Cretaceous radiolarian biostratigraphy of the Urakawa and Obira areas, Hokkaido. Science Reports of the Tohoku University. Second Series, Geology. 1982;**52**: 1–75

[35] Ueda H, Miyashita S. Tectonic accretion of a subducted intraoceanic remnant arc in Cretaceous Hokkaido, Japan, and implications for evolution of the Pacific northwest. The Island Arc. 2005;**14**:582–598

[36] Van der Voo R, Spakman W, Bijwaard H. Mesozoic subducted slabs under Siberia. Nature. 1999;**397**:246–249

[37] Vissers RLM, van Hinsbergen DJJ, Meijer PT, Piccardo GB. Kinematics of Jurassic ultra-slow spreading in the Piemonte Ligurian ocean. Earth and Planetary Science Letters. 2013;**380**:138–150.

[38] Wilson M. Igneous Petrogenesis. London: Chapman & Hall; 1989. p. 466.

[39] Winchester JA, Floyd PA. Geochemical discrimination of different magma series and their differentiation products using immobile elements. Chemical Geology. 1977;**20**: 324–343

[40] Zhou XM, Li WX. Origin of Late Mesozoic igneous rocks in Southeastern China: Implications for lithosphere subduction and underplating of mafic magmas. Tectonophysics. 2000; **326**:269–287

[41] Zonenshain LP, Kuzmin MI, Natapovm LM. Geology of the USSR: A Plate-Tectonic Synthesis. Washington, DC: AGU; 1990. p. 242.

Chapter 4

Cretaceous Research: A New Constraint on the Paleostress Regime of Southwest Japan Based on Microfabric Analysis of a Granitic Pluton

Yasuto Itoh

Additional information is available at the end of the chapter

http://dx.doi.org/10.5772/67357

Abstract

Origin of a conspicuous microscopic fabric of a granitic pluton in the eastern part of southwest Japan, which has been an autochthonous component of the East Asian margin since the late Mesozoic, was elucidated through well-organized rock magnetic and geochronological investigations. The voluminous Late Cretaceous Toki Granite suffered a thermochemical event succeeding to its emplacement, which resulted in tightly grouped principal axes of the anisotropy of magnetic susceptibility (AMS). The magnetic fabrics originate from ferromagnetic and paramagnetic minerals precipitated upon the surface of fractures developed under regional tectonic stress. Together with previous studies of three-dimensional microcrack generation and intrusion sequence of dike swarms in the Toki Granite, temporal changes in the azimuths of AMS principal axes delineate a drastic shift in tectonic stress regime along the Asian continental margin from the Late Cretaceous to the Paleocene. This remarkable event is probably linked to a regional unconformity contemporaneously formed on the convergent margin.

Keywords: paleostress, rock magnetism, anisotropy of magnetic susceptibility, fracture, tectonics

1. Introduction

For years, measurements of the anisotropy of magnetic susceptibility (AMS) have been applied to granitic rocks to determine their primary microfabric, a characteristic that reflects the azimuth of their original viscous emplacement flow (e.g., [1]). Their incompetence to strain means in general that the original fabrics are readily preserved [2]. On the other hand, AMS fabric has often been used to indicate the orientation of fractures developed after the initial emplacement of granitic plutons. For example, Itoh and Amano [3] executed an elaborate rock magnetic study on a plutonic body and found that granite samples adjacent to a large fault are remagnetized, and that their anisotropy forms an oblate shape nearly parallel to the azimuth of the fault plane. Based on microscopic observation, they confirmed the presence of fine authigenic particles of iron oxides within fault-related veins, which undoubtedly contribute to the conspicuous AMS fabric. Another notable point among their paleomagnetic results is that nested blocks of plutons had suffered differential rotation bounded by numerous orthogonal fractures.

Uplift and cooling of a granitoid body result, in general, in fracture formation influenced by intragranular microscale cracking of quartz grains [4], which develops because quartz undergoes stronger thermal contraction than feldspar. The regional stress field regime controls the formation of intrinsic microcracks whose orientations are normal to the orthogonal principal stress axes. It is conceivable that the network of cracks normal to the minimum axis is the most open and permeable pathway of formation fluids. Thus, the AMS fabric, which originates from ferromagnetic minerals precipitating on the fracture surface, mimics the paleostress field. The author attempts in this case study to verify the reliability of AMS data as a stress indicator by analyzing underground samples free from secondary weathering effects.

2. Geological settings

Analytical samples used in the present rock magnetic study were taken from the Mizunami underground research laboratory (abbreviated as MIU; see **Figure 1**) of the Japan Atomic Energy Agency (JAEA). It is an underground laboratory that tests the in situ stability of the upper crust, which consists, around the MIU, of a Jurassic to Cretaceous accretionary complex, Cretaceous to Paleogene volcano-plutonic complex, Miocene marine sediments and Miocene to Pliocene nonmarine clastics in ascending order [5, 6]. In the MIU's main and ventilation shafts, the base of the sedimentary units was confirmed at 166.27 m and 168.67 m below the surface, and the downside comprises a pluton of Toki Granite.

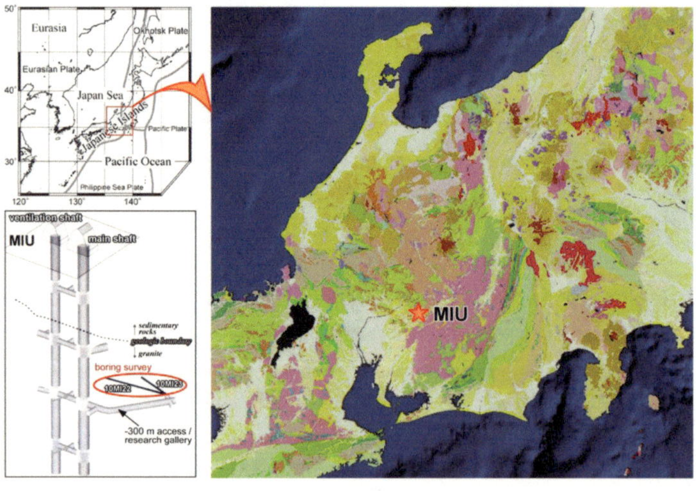

Figure 1. Index maps of the study drilling site (MIU: Mizunami underground research laboratory). The geologic map is after [5].

Toki Granite is one of the Late Cretaceous granitoids extensively distributed in southwest Japan. The pluton has yielded a CHIME monazite age of 68.3 ± 1.8 Ma [7] and whole-rock Rb-Sr ages of 72.3 ± 3.9 Ma [8] and 68.2 ± 1.1 Ma [9]. It is a stock, 14 km × 12 km in areal extent [10], having a vertical thickness of at least 1.5 km based on borehole investigations by JAEA.

3. Paleo-/rock magnetism

3.1. Sampling and methods

Paleo-/rock magnetic sampling was executed on core samples from a research borehole (10MI22) drilled horizontally from the 300-m level access/research gallery (**Figure 1**). **Figure 2** shows the lithology column and fracture density along the borehole. We obtained fresh granites from six sites (MI2201 ~ 05, 07), dark-colored aphyric intrusive rocks from two sites (MI2209, 10) and greenish altered granites from two sites (MI2206, 08). The intrusive body is closely associated with the main fault observed in the drift. Altered samples MI2206 and MI2208 are located in a fracture zone and in the alteration halo of the intrusion, respectively. Conventionally drilled core samples were successfully oriented on the basis of the correlation between fractures on the core surface and the side-wall images for the cored interval obtained by a borehole television (BTV). **Figure 3** shows typical examples of fractured core photo (a) and BTV image of the side-wall (b).

Ten to eighteen standard-sized cylindrical specimens (25 mm diameter, 22 mm long) were taken from 10 points along the Toki Granite core. Bulk initial magnetic susceptibility was measured for all the specimens using a Bartington susceptibility meter (MS2). We conducted progressive thermal demagnetization (PThD) testing on selected specimen from each site. Natural remanent magnetization (NRM) was measured using a cryogenic magnetometer (760-R SRM, 2-G Enterprises) within a magnetically shielded room at Kyoto University and spinner magnetometers (SMM-85, Natsuhara-Giken; SSM-1A, Schonstedt Instrument) at Osaka Prefecture University. The PThD test was performed, up to 680°C in air, using a non-inductively wound electric furnace with an internal residual magnetic field less than 10 nT.

3.2. Basic sample properties

For the fresh granites, we found semistable components to have a converging trend on origin of the vector-demagnetization diagrams across a broad distribution of unblocking temperatures (T_{UB}) up to 580°C after the northerly component was demagnetized at around 300°C, whereas the altered granites were highly unstable during PThD treatment, and the magnetic components of most of the specimens were not isolated. On the other hand, we successfully identified the converging trend on origin of the vector-demagnetization diagrams and isolated stable magnetic components with a broad distribution of T_{UB} up to 680°C for the intrusive rocks. The directions of the characteristic remanent magnetization (ChRM) were calculated using a three-dimensional least squares analysis technique after [11].

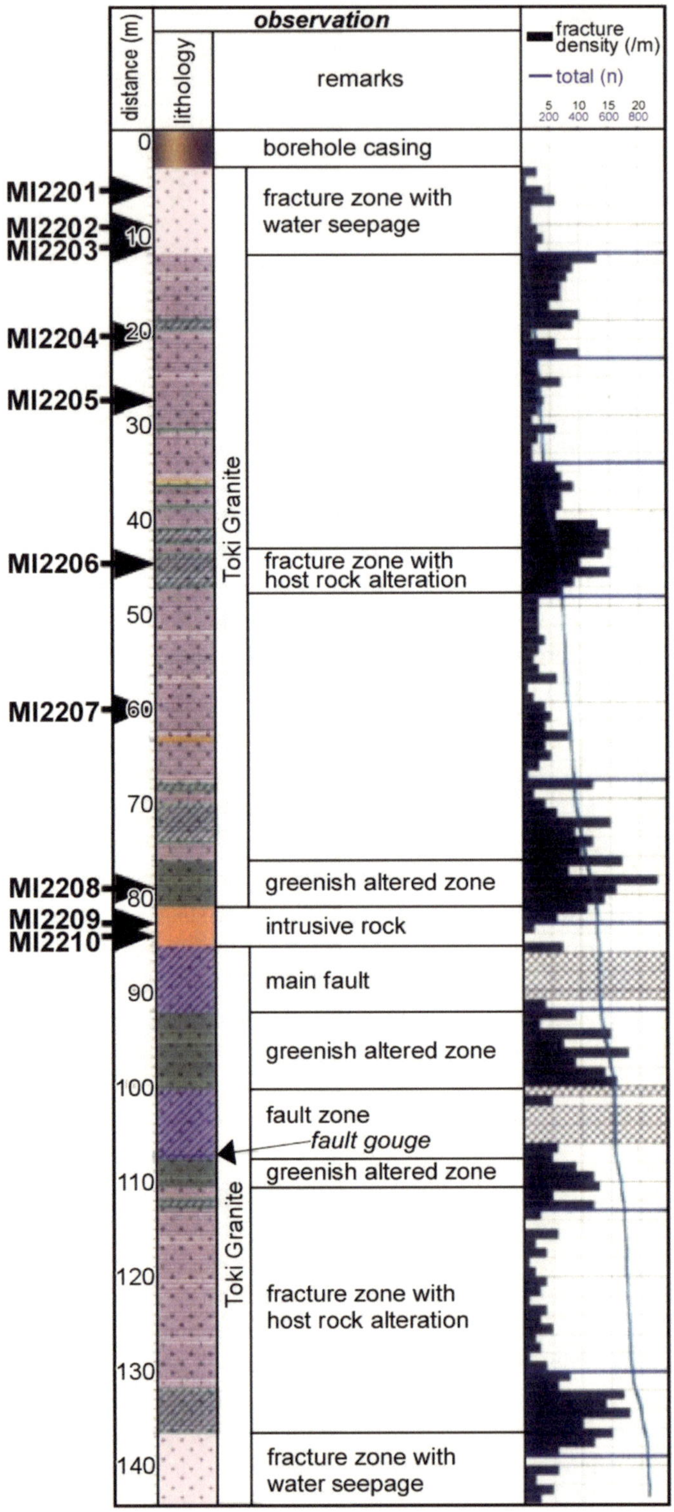

Figure 2. Columnar section of the 10MI22 borehole with rock magnetic sampling horizons.

http://dx.doi.org/10.5772/67357

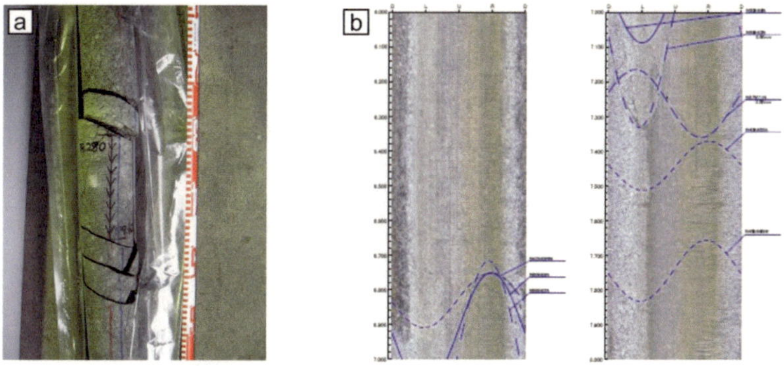

Figure 3. Determination of the orientation of conventionally drilled core samples. Fractures in a core (a) and side-wall images obtained by a borehole television (BTV; b) were correlated.

Site-mean ChRM directions in geographic coordinates are presented in **Figure 4**. As mentioned above, altered granites were thermally unstable, and number of data points was too small to obtain statistical parameters. It should be noted that the remanence directions of the fresh granites show a large scatter even in a single site. Based on thermochronological analysis, Sasao et al. [12] showed incipient rapid cooling of the Toki Granite and argued that the considerable dispersion of paleomagnetic directions in the intact rock may be attributed to plastic deformation of granites during the cooling period. The diversity of the magnetic properties of the samples may originate from thermochemical changes after the pluton was emplaced because the significantly weak NRMs of the altered rocks (**Figure 5**) imply leaching of iron oxide. Decreases in bulk initial susceptibility and increases in the frequency dependence (the F factor defined by [13]) of altered rocks (**Figure 6**) indicate a loss of magnetic carrier and smaller magnetic particles, respectively. The author attempts to identify ferromagnetic minerals in the following sections.

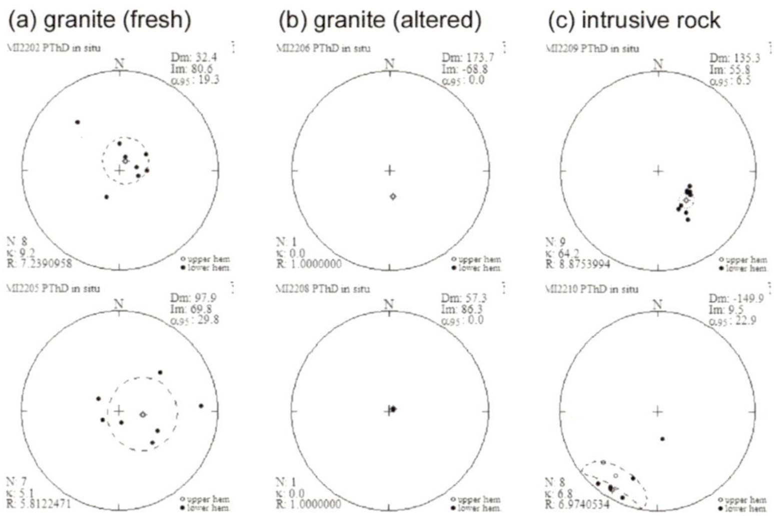

Figure 4. Equal-area projections of site-mean magnetic directions obtained from core samples of fresh granite (a), altered granite (b) and intrusive rock (c) along the 10MI22 borehole. Solid and open symbols are data plotted on the lower and upper hemispheres, respectively. Dotted ovals show 95% confidence limits.

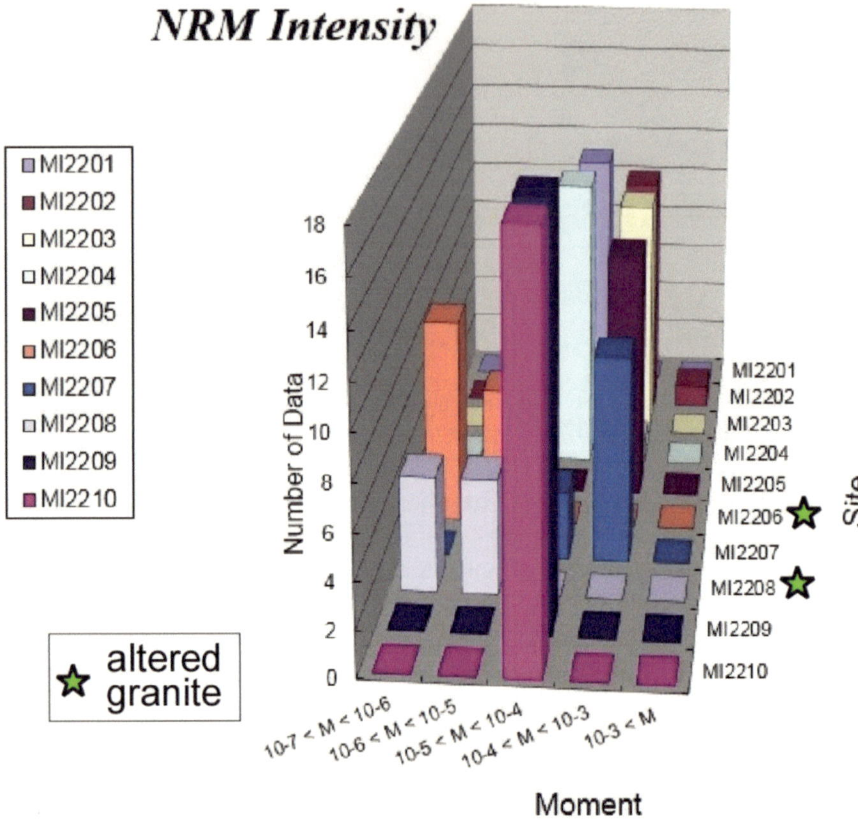

Figure 5. Comparison of intensities of natural remanent magnetization (NRM) of core samples of the 10MI22 borehole.

3.3. Hysteresis properties

Magnetic hysteresis parameters were determined with an alternating gradient magnetometer (MicroMag 2900, Princeton Measurements Corporation) at Kyoto University. Samples from 10 sites were gently ground in a mortar. From the ground material, 50 rock chips as large as 1 mm were randomly selected from each site. In addition, portions of the powder (about 50 mg) were wrapped in plastic film to prepare 50 samples per site. Because felsic minerals are more resistant to the grinding process, they tend to be concentrated in the chip samples, whereas the more fragile mafic minerals tend to concentrate in the powder samples. Among the 10 sites, magnetization of the intrusive rocks did not reach saturation at the highest field of the apparatus, so they were then excluded from the following analysis.

Figure 7 presents correlation plots of hysteresis parameters, namely, Jrs/Js versus Hcr/Hc [14], using values of the coercivity of remanence (Hcr) obtained through backfield demagnetization experiments. Trends a, b and c originate from synthetic and crushed natural magnetite [15], single-domain (SD) and multidomain (MD) mixtures of magnetite [16] and superparamagnetic (SP) and SD mixtures of magnetite [17, 18], respectively. As for the fresh granites (1), most of the data are found around mixture trends of various sizes of magnetite, implying that the effect of secondary mineral precipitation remains minimal. A small number of chip samples are plotted between trends for SD and MD mixtures of magnetite (b) and SP and SD mixtures of magnetite (c), suggesting differences in grain-size distribution. Data for the altered granite (2) fall completely separate from the SD and MD mixture trend and imply dominance of finer grains. The logarithmic parameters of chip data resemble data for remagnetized limestones, whose remanence is carried by authigenic magnetite having negligible shape anisotropy [17].

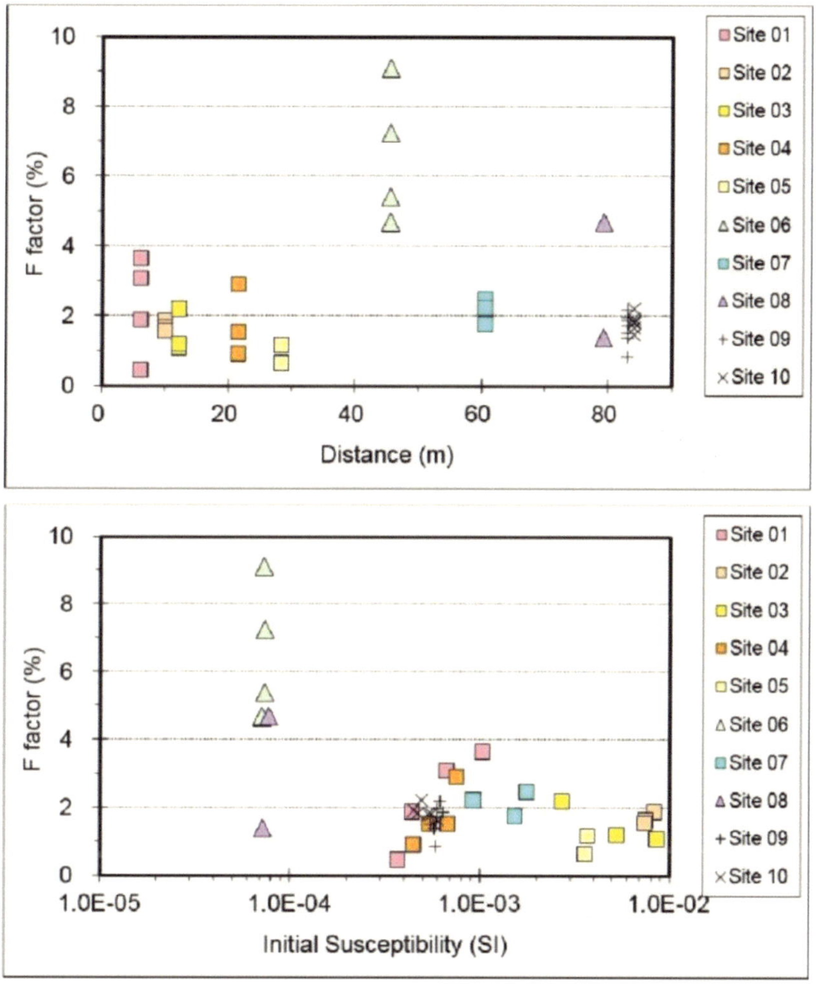

Figure 6. F-factor plots for core samples of the 10MI22 borehole. Symbol: square = fresh granite, triangle = altered granite, cross = intrusive rock.

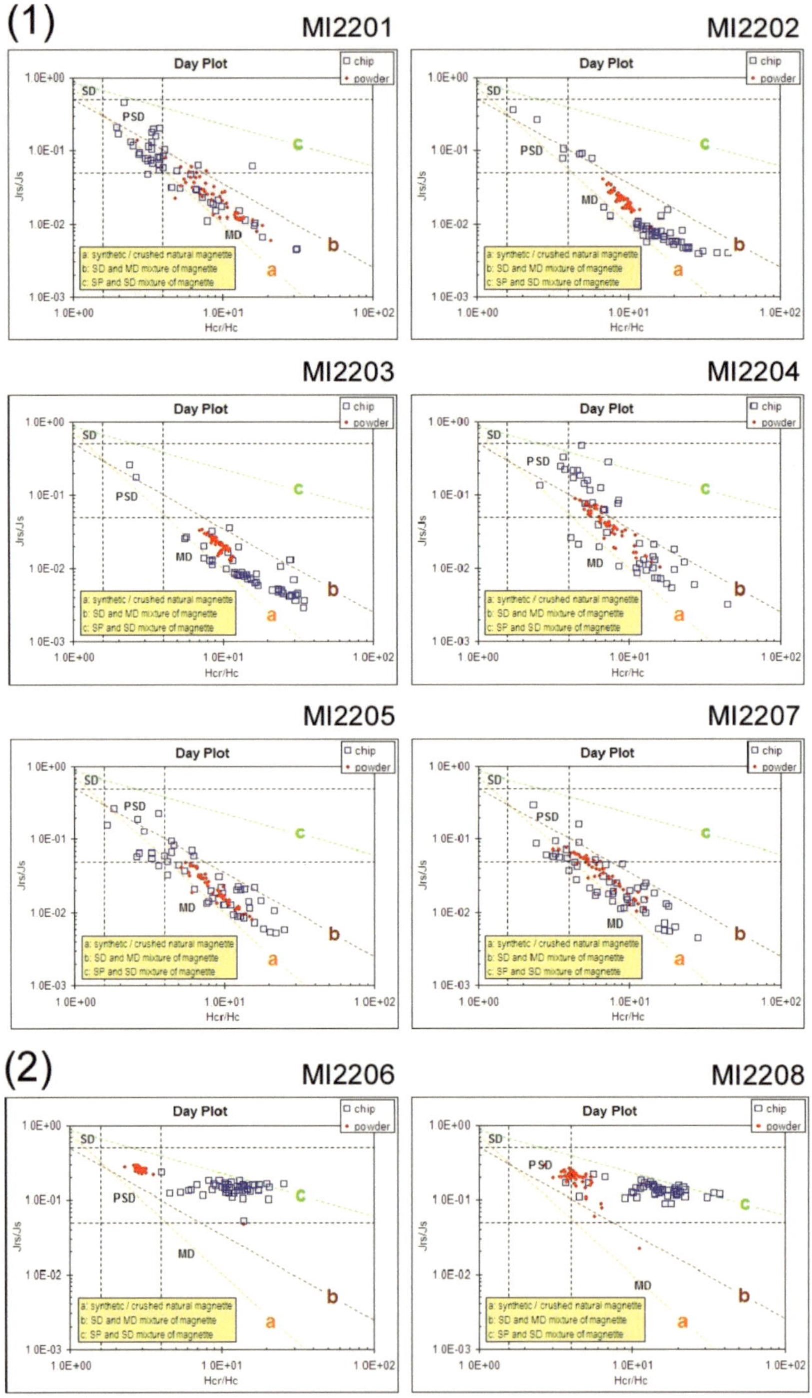

Figure 7. Logarithmic plots of hysteresis parameters [14] representing samples of fresh granite (1) and altered granite (2). See text for method of preparing chip and powder samples. Trends a, b and c originate from synthetic and crushed natural magnetite [15], single-domain (SD) and multidomain (MD) mixtures of magnetite [16] and superparamagnetic (SP) and SD mixtures of magnetite [17, 18], respectively.

Figure 8 shows examples of hysteresis loops for Toki Granite. The raw diagram seems to suggest an absence of ferromagnetic material. After correction to a linear gradient of paramagnetism, we are able to recognize a weak signature of ferromagnetic behavior. The wider loop of the altered granite (powder samples) is indicative of SD magnetite dominance. A noteworthy fact is that the altered chip sample exhibits the so-called wasp-waisted loop shape that originates from the presence of high-coercivity components related to a remagnetization event (e.g. [16]).

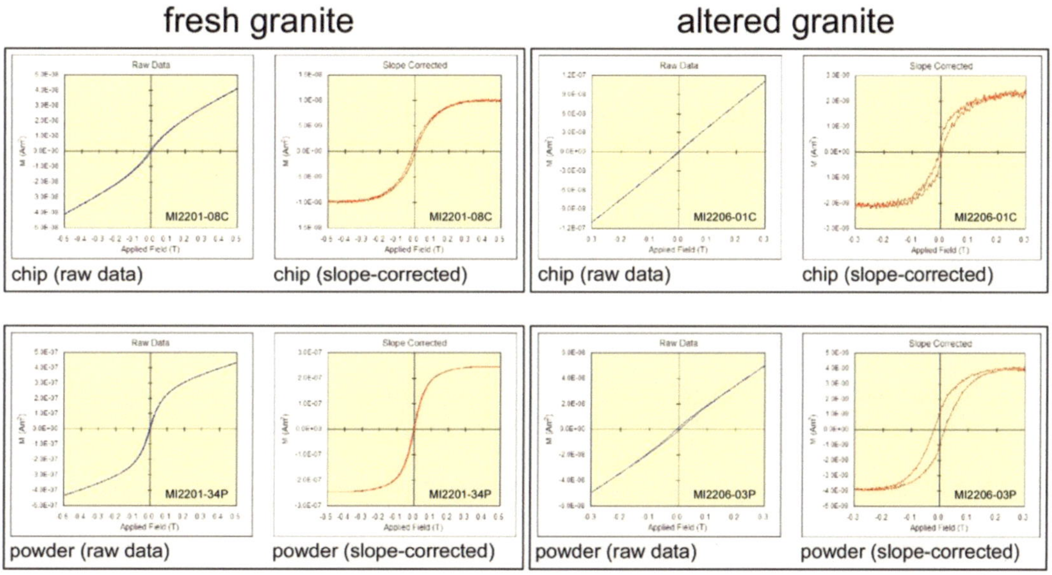

Figure 8. Examples of hysteresis loops for the Toki Granite (Blue: raw data, Red: data after correction to a linear gradient of paramagnetism). See text for preparation method of chip and powder samples.

Figure 9 shows logarithmic plots of bulk hysteresis parameters for 10 horizons of the Toki Granite measured on a vibrating sample magnetometer (VSM; MicroMag3900, Princeton Measurements Corporation) at the Center for Advanced Marine Core Research, Kochi University. Each sample, consisting of up to 1 cm^3 of crushed rocks, was contained in a gelatin capsule for the measurements. This apparatus is able to deal with bulk samples including intrusives and reconfirmed the deviated trends of the altered granites and intrusive rocks.

3.4. Isothermal remanent magnetization (IRM) experiments

3.4.1. Spectrum of coercive force

In order to identify carriers of magnetic components in the samples, isothermal remanent magnetization (IRM) experiments were undertaken. Stepwise acquisition of IRM was performed according to an analytical technique developed by [19]. **Figure 10** shows a linear acquisition plot (LAP) and gradient of acquisition plot (GAP) of the IRM progressively acquired in direct magnetic fields of up to about 3 T after pretreatment of alternating field demagnetization at 100 mT. The plots generated from a majority of the IRM data for fresh granite (a) can be matched by single magnetic components with relatively low $B_{1/2}$ values (18–25 mT; the field at which half of the IRM saturation is reached), indicating the existence of low coercive force (Hc) ferromagnetic mineral. On the basis of the T_{UB} spectra mentioned earlier, we consider that the remanent magnetization of the fresh granites resides in titanomagnetite.

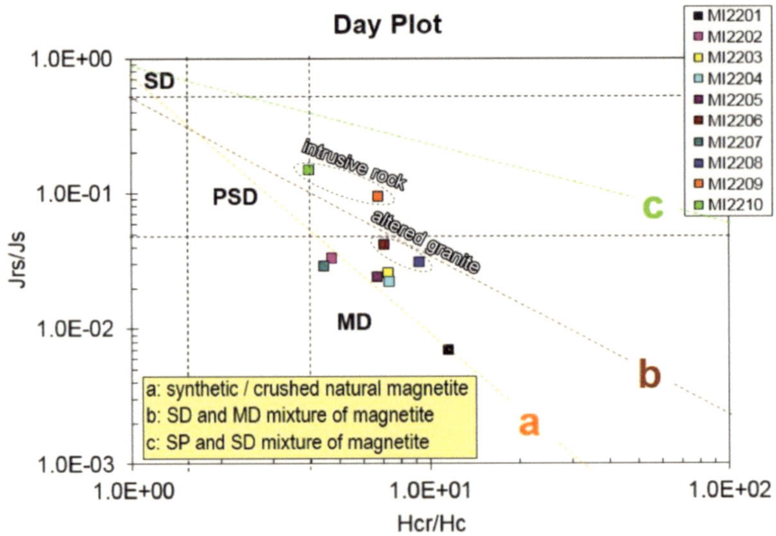

Figure 9. Logarithmic plots of bulk hysteresis parameters [14] for 10 horizons of the Toki Granite measured on a vibrating sample magnetometer (VSM). Trends a, b and c originate from synthetic and crushed natural magnetite [15], single-domain (SD) and multidomain (MD) mixtures of magnetite [16] and superparamagnetic (SP) and SD mixtures of magnetite [17, 18], respectively.

As for the altered granite (**Figure 10b**) and intrusive rock (c), the plots generated from the IRM data can be matched by two magnetic components with distinct $B_{1/2}$ values of 36–40 mT and 631–1122 mT, indicating the coexistence of low and high Hc ferromagnetic minerals. The contribution of the high Hc spectrum is overwhelming for intrusives. On the basis of the T_{UB} spectra mentioned before, we consider that the remanent magnetization of the samples that suffered a thermochemical change resides in titanomagnetite and hematite with various mixing ratios.

3.4.2. *Thermal demagnetization of orthogonal IRMs*

We executed PThD of composite IRMs for selected specimens. Based upon the procedure proposed by [20], composite IRMs were imparted by applying direct magnetic fields of 3.0, 0.4 and then 0.12 T onto the specimens in three orthogonal directions. As shown in **Figure 11**, the decay curve of the IRM components through PThD testing for fresh granite (a) indicates that the dominant magnetic phase is generally the low Hc (<0.12 T) soft fraction with a broad spectrum of T_{UB} up to 580°C. In such a case, the major carrier of the high-T_{UB} component of NRM is titanomagnetite.

http://dx.doi.org/10.5772/67357

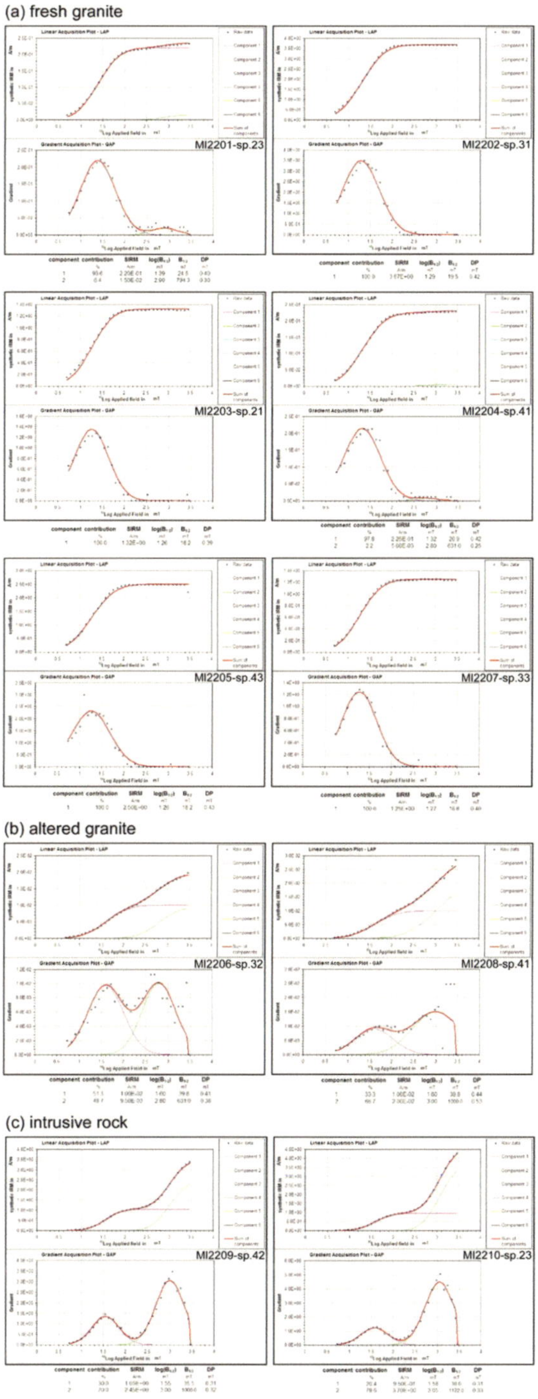

Figure 10. Linear acquisition plot (LAP) and gradient of acquisition plot (GAP) of isothermal remanent magnetization (IRM) acquired in direct magnetic fields of up to 3 T for fresh granite (a), altered granite (b) and intrusive rock (c). Specimens were processed after alternating field demagnetization at 100 mT.

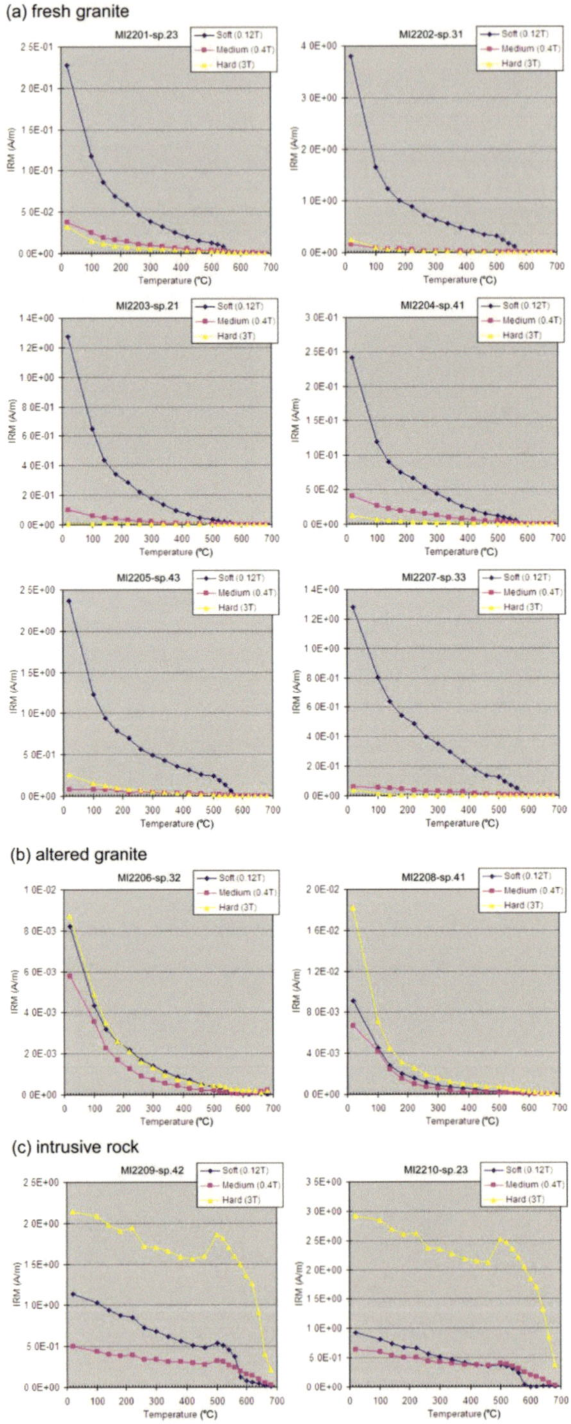

Figure 11. Thermal demagnetization curves of three orthogonal IRM components imparted on specimens from fresh granite (a), altered granite (b) and intrusive rock (c).

We identified both the medium (0.12 < Hc < 0.4 T) and hard (0.4 < Hc < 3.0 T) fractions for altered granite (b) and intrusive rock (c). In the altered granite, the medium (T_{UB} < 580°C) and hard fractions (T_{UB} < 680°C) are thought to be carried by SD-size magnetite and hematite, respectively. The medium (T_{UB} < 680°C) and hard components (T_{UB} < 680°C) of the intrusive rock probably reside in MD-size and SD-size hematite, respectively. Thus, the author has proven that the NRMs preserved in the underground samples are carried by mixture of titanomagnetite and hematite at various mixing ratios.

3.5. Magnetic susceptibility fabrics

Figure 12 presents the magnitudes of the magnetic fabrics in core samples with various lithologies. Obviously, thermochemical change results in remarkable decreases in the degree of anisotropy (P_J). It is also noteworthy that the shape parameter (T) values of the altered granites and intrusive rocks are in positive domain without exception, suggesting an oblate AMS fabric.

Figure 13 delineates the site-mean AMS fabrics of (a) fresh Toki Granite and (b) altered granite and intrusive rock. The fabric of the intact pluton has quite a large scatter, and no specific directional trend is observed. In sharp contrast with this, the principal AMS axes of the altered granites and intrusive rocks are tightly clustered with similar orientations. Considering the very weak ferromagnetic signature in the raw hysteresis loops (see Section 3.3), the fabric would appear to be carried by both authigenic iron oxides and platy grains of iron-bearing silicate minerals such as biotite or chlorite, whose shape anisotropy enhances the degree of AMS.

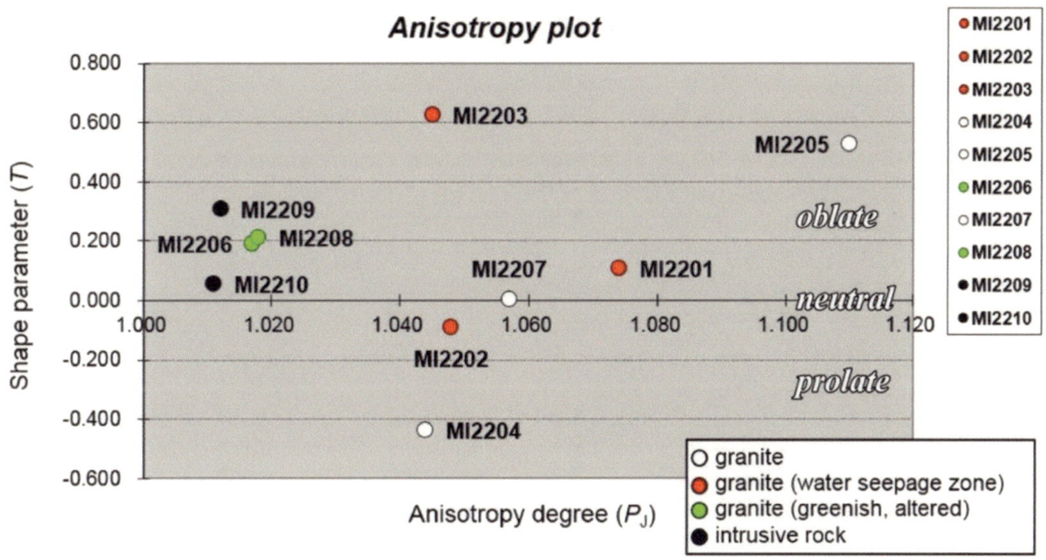

Figure 12. Magnitudes of magnetic fabrics in core samples with various lithologies.

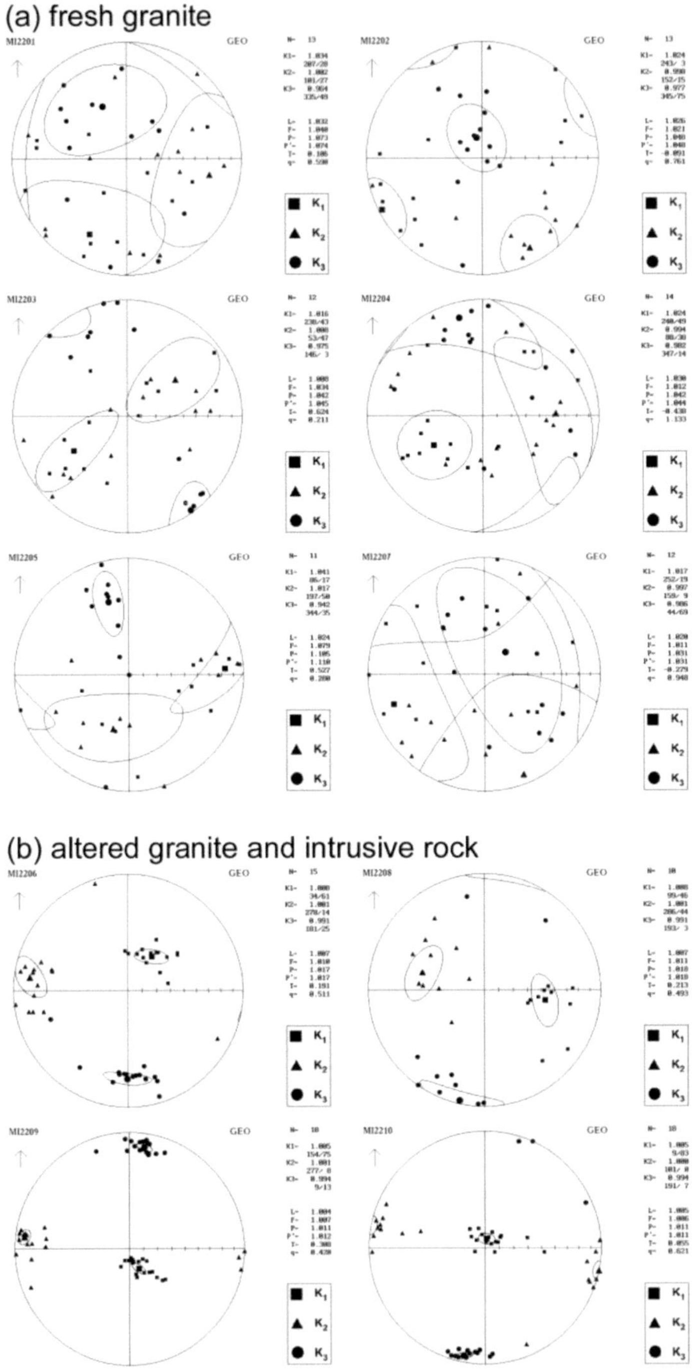

Figure 13. Anisotropy of magnetic susceptibility (AMS) fabric (principal susceptibility axes) for all specimens from the core sampling sites of the Toki Granite plotted on the lower hemisphere of equal-area projections. Data are shown in geographic coordinates. Ovals surrounding the mean directions of the three axes (shown by larger symbols) are 95% confidence regions. AMS data are presented in two groups, fresh granite (a) and altered granite and intrusive rock (b), which show similar fabrics.

4. Temporal transition of tectonic stress

Our rock magnetic experiments suggested a conspicuous AMS fabric in the altered granites and intrusive rocks. For intrusive rocks, magnetic anisotropy is generally related to an intrusion process, namely, the maximum and intermediate axes are bound to the flow plane of the magma [2]. On the assumption that the AMS-carrying ferromagnetic and paramagnetic minerals in the altered granites were precipitated and aligned on the surface of inherently orthogonal cooling fractures in the granites [4], the magnetic anisotropy can act as a proxy for the dike swarm that indicates paleostress field. As the authigenic minerals tend to grow on the most permeable open fracture network, the foliation plane of the anisotropy ellipsoid, which is perpendicular to the minimum axis (K_3), is expected to be parallel to dike elongation. In the next section, the author reviews previous studies of tectonic stress around the MIU site and tests the usability of the AMS data as a stress indicator.

4.1. Previous studies

Figure 14a presents orientation contour diagrams of healed microcracks (HCs), sealed microcracks (SCs) and open microcracks (OCs) in the Toki Granite for surface outcrops (upper) and the DH-15 borehole (lower) after [21]. They considered that the north-south trending HC data reflect the regional paleostress field, and assumed that they were formed around 60 Ma based on K-Ar ages of the pluton and the formation temperature (ca. 300–400°C) of fluid inclusions estimated from microthermometry for intrusion depth (3.5 km = 100 MPa). Later high-angle fracture sets of the SCs and OCs show quite a different trend and imply a remarkable change in the stress regime. The youngest OC was thought to be formed in the Miocene based on the exhumation and burial history of the granite, whereas the origin of the SCs filled with carbonate minerals was not clearly discussed.

In igneous terranes, paleostress study is often based on the azimuth of a dike swarm, which coincides with the σH_{max} direction at the time of intrusion [23]. Near the MIU site, there are two conspicuous sets of dikes within the Toki Granite. Hoshi and Nishimura [22] described the granitic body as having been first intruded by north-south trending aplite dikes and then cut by a high-magnesium andesite dike swarm with an east-west trend. As shown in **Figure 14b**, the three-dimensional orientations of the aplite and andesite dikes show a striking resemblance to those of the HC and SC/OC data of [21]. Their cross-cutting relationship is obvious in **Figure 14c**. Oikawa et al. [24] reported a K-Ar age of the high-magnesium andesite at 59.6 ± 1.4 Ma. This points to a drastic change in regional stress that may have been caused by a tectonic event around the end of the Cretaceous.

4.2. Radiometric dating

To consider the time sequence of stress changes envisaged from dike orientations, placing the timing of the aplite intrusion between the emplacement of the granitic pluton and the anomalous activity of the high-magnesium andesite is of great importance. Therefore, the author executed radiometric dating on the igneous rock of unknown age.

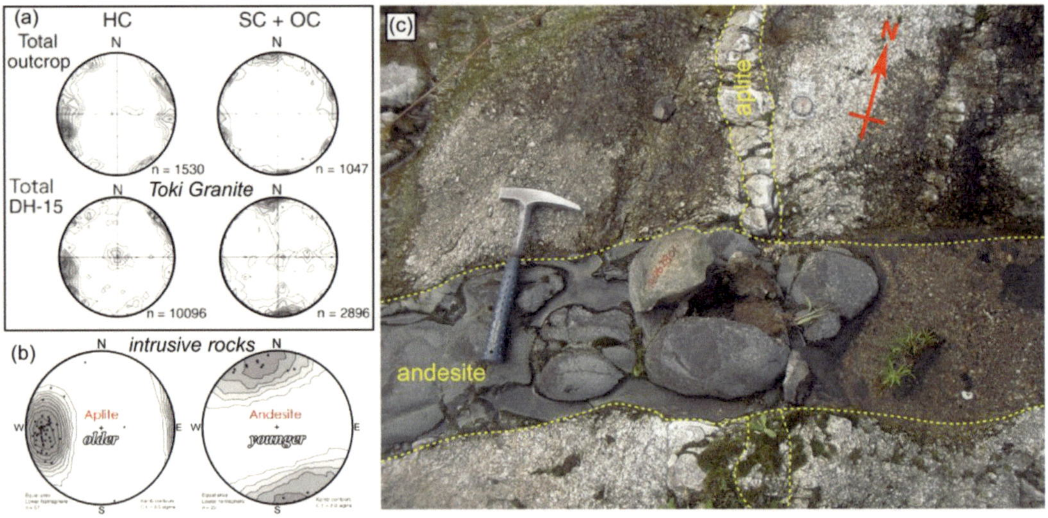

Figure 14. (a) Orientation contour diagrams of healed microcracks (HCs), sealed microcracks (SCs) and open microcracks (OCs) in the Toki Granite for surface outcrops (upper) and the DH-15 borehole (lower) after [21]. (b) Poles of aplite dikes (left) and andesite dikes (right) intruded in the Toki Granite after [22]. (c) Photo of an outcrop showing cross-cutting relationship between intrusive rocks hosted by the Toki Granite.

4.2.1. Samples

Dating samples were taken from an outcrop (Lat. = 35.3522°N, Long. = 137.1715°E) by the Toki River running through exposures of the Cretaceous granites. Samples 14062801 and 14062802 were obtained from the andesitic dike and the aplite, respectively. Leucocratic aplite (Sp. 14062802) yielded affluent zircon grains during mineral separation, whereas the andesite lacked them. Hence, we concentrated on the aplite sample.

4.2.2. Methods

The fission track (FT) dating method is after [25]. U-Pb age data were obtained using inductively coupled plasma-mass spectrometry (ICP-MS) combined with an excimer laser ablation (LA) sample introduction system. The U-Pb age determinations on zircon samples were performed after chemical leaching using 47% HF for 20 hours at room temperature or after FT etching using a KOH-NaOH eutectic solution for 22 hours at 225°C.

4.2.3. Results

The results of the FT dating of the zircon grains are summarized in **Table 1a** and **Figure 15a**. Analysis was executed on a total of 15 crystals and provided a weighted mean age of 53.2 ± 2.6

Ma. The results of a chi-squared test were negative because of considerable variation of uranium concentration as anticipated from the samples' strong zonal arrangement.

No.	Ns	Nu	S × 10^{-5} (cm^2)	ρ_s × 10^{-7} (cm^{-2})	ρ_u × 10^{-9} (cm^{-2})	Ns/Nu × 10^3	T (Ma)	σ_T (Ma)	U (ppm)
14	26	457	0.20	1.30	2.29	56.89	99.48	20.36	250
11	83	1631	1.00	0.83	1.63	50.89	89.06	10.50	178
15	16	317	0.40	0.40	0.79	50.47	88.33	22.85	87
2	73	1461	1.20	0.61	1.22	49.97	87.45	10.93	133
7	36	751	0.60	0.60	1.25	47.94	83.92	14.62	137
4	55	1227	0.80	0.69	1.53	44.82	78.51	11.17	168
9	99	2656	1.20	0.83	2.21	37.27	65.35	7.07	242
12	86	2401	1.00	0.86	2.40	35.82	62.81	7.24	262
6	79	2848	1.00	0.79	2.85	27.74	48.70	5.81	311
13	59	2266	0.60	0.98	3.78	26.04	45.72	6.24	413
1	62	2524	0.60	1.03	4.21	24.56	43.14	5.75	460
3	31	1484	0.60	0.52	2.47	20.89	36.71	6.79	270
8	29	1420	1.20	0.24	1.18	20.42	35.89	6.85	129
10	56	2840	1.20	0.47	2.37	19.72	34.65	4.83	259
5	61	3771	0.90	0.68	4.19	16.18	28.44	3.81	458

Ns is number of spontaneous tracks, Nu is number of ^{238}U counts, S is analyzed area of crystal, ϱ_s is density of spontaneous tracks, ϱ_u is density of ^{238}U counts, σ_T is error for each grain age (1 σ), U is uranium density. Uranium concentration for standardization using 91,500 standard zircon is 0.864 × 10^6 cm^2. Epsilon (ε) corresponding to conventional zeta (ζ) value for Fish Canyon Tuff zircon is 40.8 ± 1.4.

Table 1a. Fission-track grain ages of zircons obtained from Sp. 14062802.

Figure 15b compares FT and U-Pb ages for zircons obtained from the aplite. The closing temperature for U-Pb ages is about 900°C in a laboratory or >700°C in natural conditions, whereas that for zircon FT ages is 240 ~ 250°C for heating over a period of 1 million years. Reflecting the difference, the U-Pb ages tend to be older with smaller scatter than the FT ages.

Results of the U-Pb dating for the zircon grains are summarized in **Table 1b** (all analyzed grains) and **Table 1c** (adopted grains). Based on the concordia plot for U-Pb zircon ages in **Figure 15c**, we excluded zircon grains with discordant data then adopted concordant grains to calculate the average. The final result is presented in **Figure 15d**.

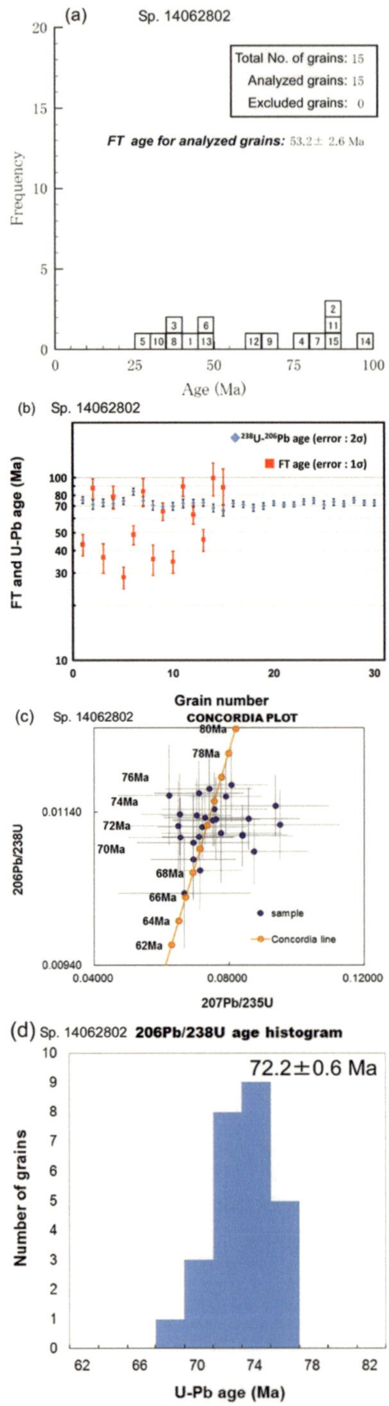

Figure 15. Radiometric dating of an aplite dike intruded in the Toki Granite (Sp. 14062802; Lat. = 35.3522°N, Long. = 137.1715°E). (a) Results of fission-track (FT) dating of zircons. (b) FT and U-Pb age plots for zircons. (c) Concordia plot for U-Pb ages of zircons. (d) Result of U-Pb dating of zircons.

http://dx.doi.org/10.5772/67357

Sample name		Total count					Isotopic ratios							Disc.*	Age (Ma)			
		^{206}Pb	^{207}Pb	^{232}Th	^{235}U	^{238}U	Th/U	$^{207}Pb/^{206}Pb$	Error 2σ	$^{206}Pb/^{238}U$	Error 2σ	$^{207}Pb/^{235}U$	Error 2σ		$^{206}Pb/^{238}U$	Error 2σ	$^{207}Pb/^{235}U$	Error 2σ
14062802	no.1	6472	327	698999	5389	742999	1.10	0.0505	±0.0045	0.01176	±0.000491	0.0808	±0.0109	concordant	75.4	±3.2	78.9	±11.0
14062802	no.2	1772	91	99999	1559	214999	0.55	0.0514	±0.0046	0.01113	±0.000649	0.0777	±0.0176	concordant	71.3	±4.2	75.9	±17.8
14062802	no.3	3682	156	158999	3169	436999	0.43	0.0424	±0.0038	0.01138	±0.000536	0.0655	±0.0117	concordant	72.9	±3.5	64.4	±11.8
14062802	no.4	2252	96	136999	1965	270999	0.59	0.0426	±0.0038	0.01122	±0.000606	0.0650	±0.0144	concordant	71.9	±3.9	63.9	±14.5
14062802	no.5	6362	319	314999	5367	739999	0.50	0.0501	±0.0045	0.01161	±0.000486	0.0791	±0.0107	concordant	74.4	±3.1	77.3	±10.8
14062802	no.6	4872	356	403999	3648	502999	0.94	0.0731	±0.0065	0.01308	±0.000578	0.1299	±0.0171	−23	83.8	±3.7	124.0	±17.3
14062802	no.7	1902	75	112999	1603	220999	0.60	0.0394	±0.0035	0.01162	±0.000662	0.0623	±0.0154	concordant	74.5	±4.3	61.3	±15.5
14062802	no.8	1702	79	138999	1516	208999	0.78	0.0464	±0.0041	0.01100	±0.000650	0.0694	±0.0168	concordant	70.5	±4.2	68.1	±16.9
14062802	no.9	3122	148	379999	2836	390999	1.14	0.0474	±0.0042	0.01078	±0.000530	0.0695	±0.0127	concordant	69.1	±3.4	68.2	±12.8
14062802	no.10	3362	238	359999	3032	417999	1.01	0.0708	±0.0063	0.01086	±0.000523	0.1045	±0.0159	−17	69.6	±3.4	100.9	±16.0
14062802	no.11	2402	77	168999	2089	287999	0.69	0.0321	±0.0028	0.01126	±0.000596	0.0491	±0.0119	16	72.2	±3.8	48.6	±12.0
14062802	no.12	3552	176	150999	3075	423999	0.42	0.0495	±0.0044	0.01131	±0.000537	0.0762	±0.0130	concordant	72.5	±3.5	74.5	±13.1
14062802	no.13	5612	256	362999	4838	666999	0.64	0.0456	±0.0041	0.01136	±0.000487	0.0704	±0.0103	concordant	72.8	±3.1	69.1	±10.4
14062802	no.14	3182	157	103999	2930	403999	0.30	0.0493	±0.0044	0.01063	±0.000520	0.0713	±0.0127	concordant	68.2	±3.3	69.9	±12.9
14062802	no.15	1072	51	90999	1015	139999	0.76	0.0476	±0.0042	0.01034	±0.000722	0.0668	±0.0198	concordant	66.3	±4.7	65.7	±19.9
14062802	no.16	3036	188	157998	2654	365999	0.51	0.0619	±0.0068	0.01124	±0.000456	0.0950	±0.0178	concordant	72.0	±2.9	92.1	±17.9
14062802	no.17	3946	219	166998	3496	481999	0.41	0.0555	±0.0061	0.01109	±0.000406	0.0840	±0.0149	concordant	71.1	±2.6	81.9	±15.1
14062802	no.18	2606	166	147998	2401	330999	0.52	0.0637	±0.0070	0.01067	±0.000461	0.0927	±0.0181	−1	68.4	±3.0	90.0	±18.2
14062802	no.19	3246	191	150998	2930	403999	0.44	0.0588	±0.0065	0.01089	±0.000430	0.0874	±0.0162	concordant	69.8	±2.8	85.1	±16.4
14062802	no.20	7076	345	700998	6158	848999	0.97	0.0488	±0.0054	0.01129	±0.000337	0.0751	±0.0117	concordant	72.4	±2.2	73.6	±11.9
14062802	no.21	8256	388	389998	7325	1009999	0.45	0.0470	±0.0052	0.01107	±0.000315	0.0710	±0.0108	concordant	71.0	±2.0	69.7	±10.9
14062802	no.22	6436	357	734998	5693	784999	1.10	0.0555	±0.0061	0.01111	±0.000342	0.0841	±0.0131	concordant	71.2	±2.2	82.0	±13.2

Sample name		Total count					Isotopic ratios							Disc.*	Age (Ma)			
		^{206}Pb	^{207}Pb	^{232}Th	^{235}U	^{238}U	Th/U	$^{207}Pb/^{206}Pb$	Error 2σ	$^{206}Pb/^{238}U$	Error 2σ	$^{207}Pb/^{235}U$	Error 2σ		$^{206}Pb/^{238}U$	Error 2σ	$^{207}Pb/^{235}U$	Error 2σ
14062802	no.23	4416	264	443998	3779	520999	1.00	0.0598	±0.0066	0.01148	±0.000403	0.0937	±0.0158	concordant	73.6	±2.6	90.9	±15.9
14062802	no.24	4136	185	172998	3489	480999	0.42	0.0447	±0.0049	0.01165	±0.000419	0.0711	±0.0133	concordant	74.7	±2.7	69.7	±13.4
14062802	no.25	2166	94	126998	1922	264999	0.56	0.0434	±0.0048	0.01107	±0.000517	0.0656	±0.0156	concordant	71.0	±3.3	64.5	±15.8
14062802	no.26	2786	135	148998	2393	329999	0.53	0.0485	±0.0053	0.01144	±0.000481	0.0756	±0.0158	concordant	73.3	±3.1	74.0	±15.9
14062802	no.27	4096	193	294998	3590	494999	0.70	0.0471	±0.0052	0.01121	±0.000405	0.0721	±0.0133	concordant	71.9	±2.6	70.7	±13.4
14062802	no.28	5046	234	264998	4236	583999	0.53	0.0464	±0.0051	0.01171	±0.000392	0.0741	±0.0129	concordant	75.0	±2.5	72.6	±13.0
14062802	no.29	5646	266	328998	4896	674999	0.57	0.0471	±0.0052	0.01133	±0.000364	0.0729	±0.0122	concordant	72.6	±2.3	71.4	±12.3
14062802	no.30	3976	221	217998	3452	475999	0.54	0.0556	±0.0061	0.01132	±0.000413	0.0858	±0.0152	concordant	72.5	±2.7	83.6	±15.3
Average ± 2SD															72.2	±6.1	76.1	±28.0
Weighted average															72.2	±0.5	74.3	±2.5

Th/U is ratio of thorium and uranium concentrations. Disc.* is degree of discordance. Positive and negative values mean left- and right-side significant offsets from concordia line, respectively.

Table 1b. U-Pb ages for all analyzed zircon grains obtained from Sp. 14062802.

Sample name		Total count					Isotope ratios							Disc.*	Age (Ma)			
		^{206}Pb	^{207}Pb	^{232}Th	^{235}U	^{238}U	$\frac{Th}{U}$	$\frac{^{207}Pb}{^{206}Pb}$	Error 2σ	$\frac{^{206}Pb}{^{238}U}$	Error 2σ	$\frac{^{207}Pb}{^{235}U}$	Error 2σ		$\frac{^{206}Pb}{^{238}U}$	Error 2σ	$\frac{^{207}Pb}{^{235}U}$	Error 2σ
14062802	no.1	6472	327	698999	5389	742999	1.10	0.0505	±0.0045	0.01176	±0.000491	0.0808	±0.0109	concordant	75.4	±3.2	78.9	±11.0
14062802	no.2	1772	91	99999	1559	214999	0.55	0.0514	±0.0046	0.01113	±0.000649	0.0777	±0.0176	concordant	71.3	±4.2	75.9	±17.8
14062802	no.3	3682	156	158999	3169	436999	0.43	0.0424	±0.0038	0.01138	±0.000536	0.0655	±0.0117	concordant	72.9	±3.5	64.4	±11.8
14062802	no.4	2252	96	136999	1965	270999	0.59	0.0426	±0.0038	0.01122	±0.000606	0.0650	±0.0144	concordant	71.9	±3.9	63.9	±14.5
14062802	no.5	6362	319	314999	5367	739999	0.50	0.0501	±0.0045	0.01161	±0.000486	0.0791	±0.0107	concordant	74.4	±3.1	77.3	±10.8
14062802	no.7	1902	75	112999	1603	220999	0.60	0.0394	±0.0035	0.01162	±0.000662	0.0623	±0.0154	concordant	74.5	±4.3	61.3	±15.5
14062802	no.8	1702	79	138999	1516	208999	0.78	0.0464	±0.0041	0.01100	±0.000650	0.0694	±0.0168	concordant	70.5	±4.2	68.1	±16.9
14062802	no.9	3122	148	379999	2836	390999	1.14	0.0474	±0.0042	0.01078	±0.000530	0.0695	±0.0127	concordant	69.1	±3.4	68.2	±12.8
14062802	no.12	3552	176	150999	3075	423999	0.42	0.0495	±0.0044	0.01131	±0.000537	0.0762	±0.0130	concordant	72.5	±3.5	74.5	±13.1
14062802	no.13	5612	256	362999	4838	666999	0.64	0.0456	±0.0041	0.01136	±0.000487	0.0704	±0.0103	concordant	72.8	±3.1	69.1	±10.4
14062802	no.14	3182	157	103999	2930	403999	0.30	0.0493	±0.0044	0.01063	±0.000520	0.0713	±0.0127	concordant	68.2	±3.3	69.9	±12.9
14062802	no.15	1072	51	90999	1015	139999	0.76	0.0476	±0.0042	0.01034	±0.000722	0.0668	±0.0198	concordant	66.3	±4.7	65.7	±19.9
14062802	no.16	3036	188	157998	2654	365999	0.51	0.0619	±0.0068	0.01124	±0.000456	0.0950	±0.0178	concordant	72.0	±2.9	92.1	±17.9
14062802	no.17	3946	219	166998	3496	481999	0.41	0.0555	±0.0061	0.01109	±0.000406	0.0840	±0.0149	concordant	71.1	±2.6	81.9	±15.1
14062802	no.19	3246	191	150998	2930	403999	0.44	0.0588	±0.0065	0.01089	±0.000430	0.0874	±0.0162	concordant	69.8	±2.8	85.1	±16.4
14062802	no.20	7076	345	700998	6158	848999	0.97	0.0488	±0.0054	0.01129	±0.000337	0.0751	±0.0117	concordant	72.4	±2.2	73.6	±11.9
14062802	no.21	8256	388	389998	7325	1009999	0.45	0.0470	±0.0052	0.01107	±0.000315	0.0710	±0.0108	concordant	71.0	±2.0	69.7	±10.9
14062802	no.22	6436	357	734998	5693	784999	1.10	0.0555	±0.0061	0.01111	±0.000342	0.0841	±0.0131	concordant	71.2	±2.2	82.0	±13.2
14062802	no.23	4416	264	443998	3779	520999	1.00	0.0598	±0.0066	0.01148	±0.000403	0.0937	±0.0158	concordant	73.6	±2.6	90.9	±15.9
14062802	no.24	4136	185	172998	3489	480999	0.42	0.0447	±0.0049	0.01165	±0.000419	0.0711	±0.0133	concordant	74.7	±2.7	69.7	±13.4
14062802	no.25	2166	94	126998	1922	264999	0.56	0.0434	±0.0048	0.01107	±0.000517	0.0656	±0.0156	concordant	71.0	±3.3	64.5	±15.8
14062802	no.26	2786	135	148998	2393	329999	0.53	0.0485	±0.0053	0.01144	±0.000481	0.0756	±0.0158	concordant	73.3	±3.1	74.0	±15.9

Sample name		Total count					Isotope ratios							Disc.*	Age (Ma)			
		^{206}Pb	^{207}Pb	^{232}Th	^{235}U	^{238}U	Th/U	$^{207}Pb/^{206}Pb$	Error 2σ	$^{206}Pb/^{238}U$	Error 2σ	$^{207}Pb/^{235}U$	Error 2σ		$^{206}Pb/^{238}U$	Error 2σ	$^{207}Pb/^{235}U$	Error 2σ
14062802	no.27	4096	193	294998	3590	494999	0.70	0.0471	±0.0052	0.01121	±0.000405	0.0721	±0.0133	concordant	71.9	±2.6	70.7	±13.4
14062802	no.28	5046	234	264998	4236	583999	0.53	0.0464	±0.0051	0.01171	±0.000392	0.0741	±0.0129	concordant	75.0	±2.5	72.6	±13.0
14062802	no.29	5646	266	328998	4896	674999	0.57	0.0471	±0.0052	0.01133	±0.000364	0.0729	±0.0122	concordant	72.6	±2.3	71.4	±12.3
14062802	no.30	3976	221	217998	3452	475999	0.54	0.0556	±0.0061	0.01132	±0.000413	0.0858	±0.0152	concordant	72.5	±2.7	83.6	±15.3
Average ± 2SD															72.0	±4.2	73.8	±16.4
Weighted average															72.2	±0.6	73.3	±2.7

Th/U is ratio of thorium and uranium concentrations. Disc.* is degree of discordance. Positive and negative values mean left- and right-side significant offsets from concordia line, respectively.

Table 1c. U-Pb ages for adopted zircon grains obtained from Sp. 14062802.

http://dx.doi.org/10.5772/67357

5. Discussion

5.1. Time sequence of igneous activity

A chronological study by the author obtained a U-Pb age of 72.2 ± 0.6 Ma for the aplite dike intruded into the Toki Granite. Considering the ages reported for the host rock [7–9], the dike swarm was emplaced in the final stage of the Late Cretaceous extensive igneous activity, as Kinoshita [26] interpreted as resulting from subduction of oceanic plates' divergent margin. Oikawa et al. [24] stated that their K-Ar age for the andesitic dike provides the upper limit of the intrusion event. They also pointed out two stages of formation of high-magnesium andesite, which is suggestive of thermal input around the continental margin. Such a thermal event may be related to ridge subduction as conceived by [26] or, alternatively, with an asthenospheric injection that provoked crustal thinning and deformation of the Far East [27].

5.2. Tectonic context of stress regimes

The azimuth of the aplite dike swarm is indicative of a north-south σH_{max} direction at around the end of the Cretaceous, whereas those of the andesite dike swarm and our AMS fabric acquired during a thermochemical event in the early Paleogene imply a drastic change of σH_{max} trend into an east-west direction. We assume that the regional stress is reflected in a wrench deformation mode of the continental margin, which was controlled by lateral motions on the longstanding Median Tectonic Line as advocated by [28]. **Figure 16** presents the Late Cretaceous to Paleocene tectonic stress transition around the eastern part of southwest Japan inferred from geologic evidences. As clarified by [29], the deformation process of the forearc accretionary complex of southwest Japan implies an enigmatic shift of shearing modes: sinistral to dextral after 89 Ma and dextral to sinistral before 76 Ma. The present study seems to indicate that the older stress regime's unraveling (**Figure 16** left) is related to the sinistral stage since ca. 76 Ma based on the timing of felsic magma activity and the aplite intrusion. The tectonic context of the younger regime, confirmed by the azimuth of the anomalous andesite dike swarm and the AMS fabric of the Toki pluton, will be understood through further effort to determine the regional extent and duration of the stress state.

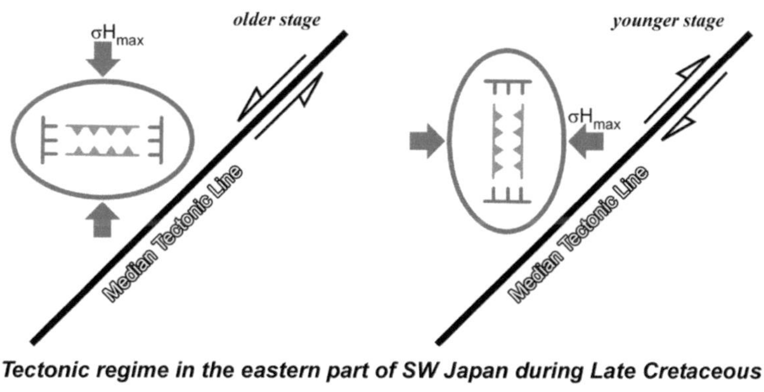

Figure 16. Late Cretaceous to Paleocene tectonic stress transition around the eastern part of southwest Japan inferred from geologic evidences presented in **Figure 14**.

6. Conclusions

Well-organized rock magnetic and geochronological investigations have revealed the origin of a unique magnetic fabric of a granitic pluton in the eastern part of southwest Japan. The Late Cretaceous Toki Granite suffered a thermochemical event, which resulted in tightly grouped principal axes of the anisotropy of magnetic susceptibility (AMS). The magnetic fabrics originate from ferromagnetic and paramagnetic minerals precipitated on fracture surfaces developed under regional tectonic stress. Together with previous studies of microcrack generation and dike intrusion in the Toki Granite, the AMS fabric delineates a drastic shift in tectonic stress state along the Asian continental margin from the Late Cretaceous to the early Paleogene.

Acknowledgements

The author is grateful to Eiji Sasao, Takashi Yuguchi and Masayuki Ishibashi for their earnest support and advice during the course of sampling at the Mizunami Underground Research Laboratory of the Japan Atomic Energy Agency and the succeeding data analysis. The author also extends his hearty appreciation to Naoto Ishikawa and Yuhji Yamamoto for the use of the rock magnetic laboratories at Kyoto University and the Center for Advanced Marine Core Research of Kochi University, respectively. Thanks are also due to Takashi Inoue for his efforts during the experiments.

Author details

Yasuto Itoh

Address all correspondence to: yasutokov@yahoo.co.jp

Graduate School of Science, Osaka Prefecture University, Osaka, Japan

References

[1] Hrouda F, Lanza R. Magnetic fabric in the Biella and Traversella stocks (Periadriatic Line): implications for the emplacement mode. Physics of the Earth and Planetary Interiors. 1989; 56: 337–348.

[2] Tarling DH, Hrouda F. The Magnetic Anisotropy of Rocks. London: Chapman & Hall; 1993. 217 p.

[3] Itoh Y, Amano K. Progressive segmentation and systematic block rotation within a plutonic body: palaeomagnetism of the Cretaceous Kurihashi granodiorite in northeast Japan. Geophysical Journal International. 2004; 157: 128–140.

[4] Vollbrecht A, Rust S, Weber K. Development of microcracks in granites during cooling and uplift: examples from the Variscan basement in NE Bavaria, Germany. Journal of Structural Geology. 1991; 13: 787–799.

[5] Geological Survey of Japan, AIST, editor. Seamless Digital Geological Map of Japan 1:200,000 (July 3, 2012 Version), Research Information Database DB084. Tsukuba: Geological Survey of Japan, AIST (National Institute of Advanced Industrial Science and Technology); 2012.

[6] Yamashita N, Kaseno Y, Itoigawa J, editors. Regional Geology of Japan – Part 5 Chubu II. Tokyo: Kyoritsu Shuppan Co.; 1988. 310 p.

[7] Suzuki K, Adachi M. Denudation history of the high T/P Ryoke metamorphic belt, southwest Japan: constraints from CHIME monazite ages of gneisses and granitoids. Journal of Metamorphic Geology. 1998; 16: 23–37.

[8] Shibata K, Ishihara S. Rb-Sr whole-rock and K-Ar mineral ages of granitic rocks in Japan. Geochemical Journal. 1979; 13: 113–119.

[9] Yuguchi T, Tsuruta T, Hama K, Nishiyama T. The spatial variation of initial $^{87}Sr/^{86}Sr$ ratios in the Toki granite, central Japan: implications for the intrusion and cooling processes of a granitic pluton. Journal of Mineralogical and Petrological Sciences. 2013; 108: 1–12.

[10] Ishihara S, Suzuki Y. Basement granites of the Toki uranium deposits in Tono region. Reports of the Geological Survey of Japan. 1969; 232: 113–127.

[11] Kirschvink JL. The least-squares line and plane and the analysis of palaeomagnetic data. Geophysical Journal of the Royal Astronomical Society. 1980; 62: 699–718.

[12] Sasao E, Yuguchi T, Itoh Y, Inoue T, Ishibashi M. Formative mechanism of inhomogeneous distribution of fractures, an example of the Toki Granite, Central Japan. In: Proceedings of 10th Asian Regional Conference of IAEG; 26–29 Sept. 2015; Uji. 6 p. (http://2015ars.com/arc/data/TP3/Tp3-P01_1075939_1493692.pdf)

[13] Heller F, Liu XM, Liu TS, Xu TC. Magnetic susceptibility of loess in China. Earth and Planetary Science Letters. 1991; 103: 301–310.

[14] Day R, Fuller M, Schmidt VA. Hysteresis properties of titanomagnetites: grain-size and compositional dependence. Physics of the Earth and Planetary Interiors. 1977; 13: 260–267.

[15] Dunlop DJ. Hysteresis properties of magnetite and their dependence on particle size: a test of pseudo-single-domain remanence models. Journal of Geophysical Research. 1986; 91: 9569–9584.

[16] Channell JET, McCabe C. Comparison of magnetic hysteresis parameters of unremagnetized and remagnetized limestones. Journal of Geophysical Research. 1994; 99: 4613–4623.

[17] Jackson M. Diagenetic sources of stable remanence in remagnetized Paleozoic cratonic carbonates: a rock magnetic study. Journal of Geophysical Research. 1990; 95: 2753–2761.

[18] Jackson M, Rochette P, Fillion G, Banerjee S, Marvin J. Rock magnetism of remagnetized Paleozoic carbonates: low-temperature behavior and susceptibility characteristics. Journal of Geophysical Research. 1993; 98: 6217–6225.

[19] Kruiver PP, Dekkers MJ, Heslop D. Quantification of magnetic coercivity components by the analysis of acquisition curves of isothermal remanent magnetisation. Earth and Planetary Science Letters. 2001; 189: 269–276.

[20] Lowrie W. Identification of ferromagnetic minerals in a rock by coercivity and unblocking temperature properties. Geophysical Research Letters. 1990; 17: 159–162.

[21] Takagi H, Miwa S, Yokomizo Y, Nishijima K, Enjoji M, Mizuno T, Amano K. Estimation of the paleostress field from the 3-D orientation distribution of microcracks and their geothermal conditions in the Toki Granite, central Japan. Journal of the Geological Society of Japan. 2008; 114: 321–335.

[22] Hoshi H, Nishimura N. Paleomagnetic results from andesite dikes in Toki, Gifu Prefecture: implications for dike emplacement and rotation. In: Abstract of Japan Geoscience Union Meeting; 20–25 May 2012; SEM21-P12; Makuhari. 2012.

[23] Nakamura K. Volcanoes as possible indicators of tectonic stress orientation—principle and proposal. Journal of Volcanology and Geothermal Research. 1977; 2: 1–16.

[24] Oikawa T, Mashima H, Tanase A, Ninomiya A, Umeda K. K-Ar age of high magnesium andesite dikes at the Toki River in Gifu Prefecture, central Japan. Journal of the Geological Society of Japan. 2006; 112: 616–619.

[25] Danhara T, Iwano H. Determination of zeta values for fission-track age calibration using thermal neutron irradiation at the JRR-3 reactor of JAEA, Japan. Journal of the Geological Society of Japan. 2009; 115: 141–145.

[26] Kinoshita O. Migration of igneous activities related to ridge subduction in Southwest Japan and the East Asian continental margin from the Mesozoic to the Paleogene. Tectonophysics. 1995; 245: 25–35.

[27] Itoh Y, Uno K, Arato H. Seismic evidence of divergent rifting and subsequent deformation in the southern Japan Sea, and a Cenozoic tectonic synthesis of the eastern Eurasian margin. Journal of Asian Earth Sciences. 2006; 27: 933–942.

[28] Itoh Y, Takemura K. Quaternary geomorphic trends within Southwest Japan: extensive wrench deformation related to transcurrent motions of the Median Tectonic Line. Tectonophysics. 1993; 227: 95–104.

[29] Tokiwa T. Timing of dextral oblique subduction along the eastern margin of the Asian continent in the Late Cretaceous: evidence from the accretionary complex of the Shimanto Belt in the Kii Peninsula, Southwest Japan. Island Arc. 2009; 18: 306–319.

Chapter 5

Tectonic Synthesis: A Plate Reconstruction Model of the NW Pacific Region Since 100 Ma

Yasuto Itoh, Osamu Takano and Reishi Takashima

Additional information is available at the end of the chapter

http://dx.doi.org/10.5772/67358

Abstract

Based on the results of interdisciplinary study from Chapters 1–4, a plate tectonic model of the northwestern Pacific region since 100 Ma is presented in this chapter. The evolution of the Pacific margin is viewed as a longstanding history of migration/amalgamation of allochthonous blocks onto the subduction zone. Such a process inevitably provoked diverse tectonic events, spatiotemporal positions of which have been discussed in this book. In order to reconcile paradoxical discrepancies in the docking process of arc fragments, the authors introduce a marginal sea plate with a spreading center that was alive in the Cretaceous. Oblique subduction of the ridge caused specific migratory igneous activity along the rim of the overriding plates, together with flips of shearing direction. Arc-trench systems on the eastern and western sides of the marginal sea plate developed following different timelines and were eventually mixed up during the plate's closure that prompted formation of a coincident Oligocene clinounconformity widespread on the Eurasian margin. Since the demise of the hypothetical plate, the tectonic regime of the northwestern Pacific margin has been controlled by the growth, namely, the rotational history and modes of convergence of the Philippine Sea Plate.

Keywords: plate motion, marginal sea, northwestern Pacific, Philippine Sea Plate, tectonics

1. Introduction

The authors' in-depth research has described a variety of tectonic events around the northwestern Pacific margin since the late Mesozoic. The architecture of the forearc basins of the Japanese archipelago depicted by high-resolution seismic profiles is indicative of a regional Oligocene unconformity related to a prevailing compressive regime (Chapter 1). A paleomagnetic investigation of the Mesozoic strata in south central Hokkaido showed significant northward transportation of crustal blocks since the Early Cretaceous and their amalgamation/collision with the continental margin with different timings by the late Paleogene (Chapter 2). Petrological and geochemical studies of the voluminous Late Jurassic to Early Cretaceous igneous rocks in Hokkaido have confirmed a typical island-arc affinity of andesitic volcanics and found signatures that suggest a collision of oceanic island arcs and seafloor spreading on the convergent margin (Chapter 3). Descriptions of microscale cracking and the magnetic fabrics of a Late Cretaceous granitic pluton, combined with an azimuthal observation of dike swarms, suggest drastic changes in the stress regimes around the Eurasian margin from the Late Cretaceous to Early Paleogene (Chapter 4).

These remarkable episodes should be closely related to the plate tectonic configuration because the northwestern Pacific has been a site of the birth and demise of oceanic plates since the late Mesozoic. As shown by Engebretson et al. [1], extremely rapid northward migration of the Izanagi Plate is key to understanding the intensive Cretaceous deformation process on plate margins. Temporal changes in the Pacific Plate's motion have controlled the formation and deformation of island arcs through the Cenozoic. Precise paleoreconstruction is difficult, however, because the Pacific Basin is surrounded by convergent margins where the evidence of geologic phenomena has been sucked deep into the interior of the earth. In this chapter, the authors attempt to construct a new kinematic model of plate motions for the northwestern Pacific region (**Figure 1**) based on our research and previous studies giving constraints on the timeline of thermal and structural events on active margins and clues to the whereabouts of allochthonous terranes since 100 Ma.

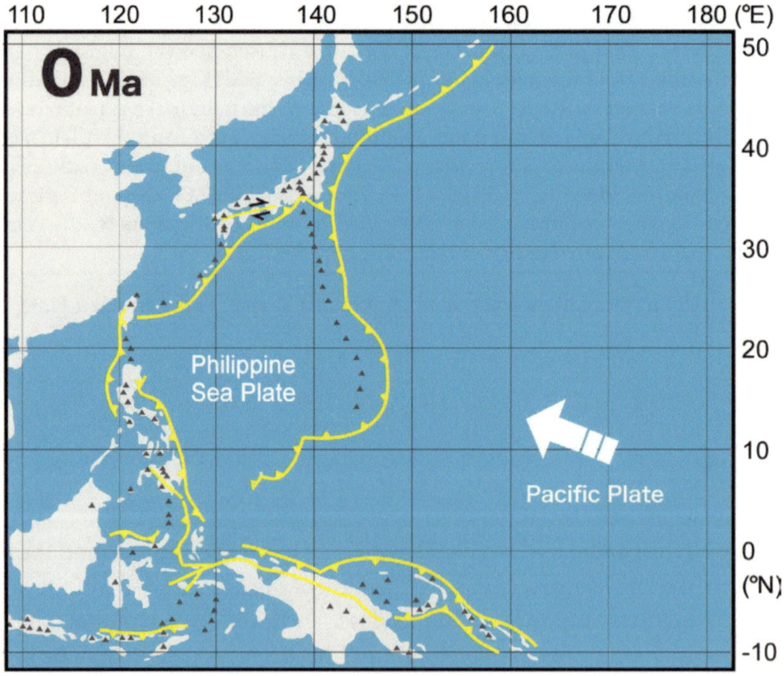

Figure 1. Present status of the plate reconstruction area of the NW Pacific region.

http://dx.doi.org/10.5772/67358

2. Constraints on plate reconstructions

We place a special emphasis on the chronicle of Cretaceous tectonic events because the ubiquitous, continuous sedimentary records of the present-day Asian continental margin are helpful in verifying our tectonic model. **Figure 2** summarizes the tectonic and/or thermal episodes on the northwestern Pacific margin throughout the period. In the following sections, highly diverse and somewhat self-contradictory event sequences are described one by one.

2.1. N-S transportation and amalgamation

In addition to the lateral motion of Hokkaido's constituents discussed in Chapter 2, significant northward migration has been reported from other areas of the Far East. Bazhenov et al. [2] did a paleomagnetic analysis of the Campanian-Maastrichtian tuffaceous siltstones and sandstones of island arc affinity from the Chayka Formation, which is distributed in south Sakhalin, and determined a paleolatitude of 27°N (**Figure 2**). They posited intra-oceanic motion of hypothetical island arcs with the Pacific Plate and their docking at the Eurasian margin at ca. 30 Ma. Another reliable paleomagnetic work focusing on Sakhalin, however, suggested that the island has a complicated structure, as does Hokkaido. Weaver et al. [3] compared their paleomagnetic inclination data with the apparent polar wander paths of major plates and concluded that a part of south Sakhalin probably evolved with the North American Plate or the Eurasian Plate. Thus, multiple origins of the northwestern Pacific landmasses have been deduced at a wide geographic scope. Northerly motion of oceanic plates also provoked intermittent development of accretionary complexes (**Figure 2**) such as the Samarkinsk terrane in Sikhote Alin [4] and the Naizawa

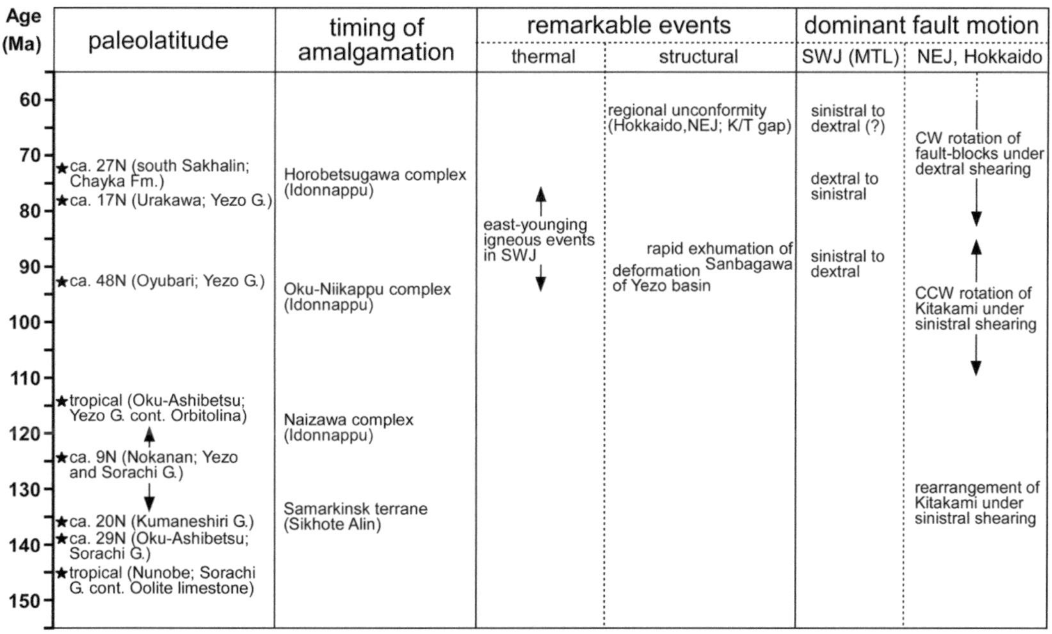

Figure 2. Tectonic and/or thermal episodes on the NW Pacific margin through the Cretaceous period.

and Horobetsugawa complexes in Hokkaido [5] in ascending order. Ueda and Miyashita [6] interpreted the Oku-Niikappu complex in the Cretaceous accretionary terranes of Hokkaido as a fragmented remnant island arc that moved with an oceanic plate and collided against the continent.

2.2. Thermal events

The eastern margin of Eurasia was a site of voluminous igneous activity in the Late Cretaceous. Radiometric ages of these rocks show a clear trend of being younger toward the northeast. Kinoshita [7] assumed that ridge subduction affected this conspicuous magmatism. He took two magmatic belts with different spatiotemporal development trends into account and attributed them to the subduction of the Farallon-Izanagi and Kula-Pacific ridges. Ridge subduction is a so-called "ace in the hole" to explain inscrutable events on convergent margins. Wallis et al. [8] reported an astonishing result that the Lu-Hf ages of eclogite samples obtained from the Sanbagawa metamorphic rocks in southwest Japan were centered at 89–88 Ma. This was identical to the Ar-Ar phengite ages of the same unit, which requires an extremely rapid exhumation of the metamorphosed terrane on the forearc. Their concept was that the approaching Izanagi-Pacific ridge triggered an anomalous uplift of the accretionary prism.

2.3. Structural events

Not only the southern part of the Cretaceous continental margin but also the northern part as well underwent intensive uplift. Tamaki et al. [9] executed one-dimensional basin modeling based on organic maturation. They investigated the vitrinite reflectance and T_{max} parameter values from Rock-Eval pyrolysis of the Yezo Group in the Oyubari area of central Hokkaido. Tamaki and Itoh [10] obtained paleomagnetic inclinations for that group, showing an affinity for a coeval paleolatitude with the North China Block [11]. They suggested notably low, constant maturation throughout the thick study section, which was confirmed by various biomarker analyses. This requires an anomalously rapid burial, probably related to thrust-stacking and/or large-scale slumping, followed by a prompt tilting/exhumation event (**Figure 2**).

In a wider geographic view, central Hokkaido has been an area of syn-accretionary exhumation. Ueda [12] demonstrated that the Early Cretaceous Iwashimizu complex in the Kamuikotan metamorphic zone, to the east of the Oyubari area, had been exhumed as a result of repetitive duplex-forming compression and extensional unroofing during the subduction of seamounts. Ando and Tomosugi [13] confirmed a younger unconformity between the Upper Maastrichtian and Upper Paleocene around the western rim of central Hokkaido and the Pacific coast of northeast Japan. The intensive deformation of the forearc probably propagated westward from the Cretaceous to the earliest Paleogene. The youngest event of the K/T unconformity is ubiquitous in the Yezo sedimentary basin. Ando [14] stated that the stacking patterns of the sequences and the timing of sequence boundaries agree with the oscillation patterns of the Haq curves, which fact led him to assume global eustasy control over the regional event.

http://dx.doi.org/10.5772/67358

2.4. Dominant sense of faulting on plate margins

As discussed in Chapter 4, the paleostress regime of southwest Japan indicates repetitive shifts in the strike-slip of the arc-bisecting Median Tectonic Line under the influence of fluctuating oceanic plate convergence from the Late Cretaceous to the early Paleogene. On the other hand, in the Early Cretaceous, northeast Japan was governed by sinistral shearing stress provoked by a rapid northward motion of the Izanagi Plate. Based on an elaborate work of structural geology in northeast Japan, Sasaki [15] determined a predeformation configuration of its forearc (Kitakami massif) restoring left-lateral displacements on arc-parallel faults (**Figure 2**).

The strong sinistral phase was replaced by a dextral phase in the Late Cretaceous. Based on the counterclockwise and then clockwise rotational sequence of a large pluton in northeast Japan revealed by paleomagnetic methods, Itoh et al. [16] argued that switching from left- to right-lateral slips on arc-parallel faults resulted in a ball-bearing clockwise motion of crustal blocks. Such a deformation mode spread over the forearc basin of northeast Japan [17] and Hokkaido and lingered through the Paleogene. This was shown by numerical modeling of pull-apart basin formation in Hokkaido [18]. Its timeline is totally discordant from that of southwest Japan. In summary, the southwestern and northeastern parts of the Japan arc seem to have belonged to different convergent margins around the Cretaceous period.

3. Reconstructions

By reviewing the Cretaceous episodes, we have recognized anew that the present-day island arcs on the northwestern Pacific margin are a hodgepodge of allochthonous and autochthonous blocks that experienced various tectonic events. For example, a long-accepted theory posits that central Hokkaido consists of a volcanic arc associated with an accretionary complex overlain by forearc basin sediments that developed somewhere on the Pacific margin during late Mesozoic era. However, the authors' paleomagnetic study (Chapter 2) revealed that the ancient "forearc" of the Sorachi-Yezo Belt contains crustal blocks derived from low latitudes, although they cannot be discriminated from in situ coeval units by clear geologic boundaries. The evolutionary process of the "backarc" of the Oshima Belt is in an even more serious turmoil. Our paleomagnetic data on the Early Cretaceous Kumaneshiri Group suggest northward transportation of the volcanic arc, whereas some of the clastics in the terrane definitely originated from the Asian craton. Kawamura et al. [19] reported SHRIMP U-Pb ages of detrital zircon grains separated from Jurassic sandstones at ca. 1800 and 2500 Ma and confirmed a Proterozoic/Archean clastic provenance for the Oshima Belt.

As shown in **Figure 3**, the authors introduce a marginal sea plate between the mother continent and the major oceanic plates to construct a model that can reconcile the varied timings of the collision/amalgamation of crustal blocks onto the continental rim and the conspicuous magmatism that is probably related to ridge subduction. Now we are ready to present a plate tectonic chronicle for the northwestern Pacific region from 100 to 10 Ma. The animations of the reconstruction (a complete set at 5-m.y. intervals) that accompany the following section are

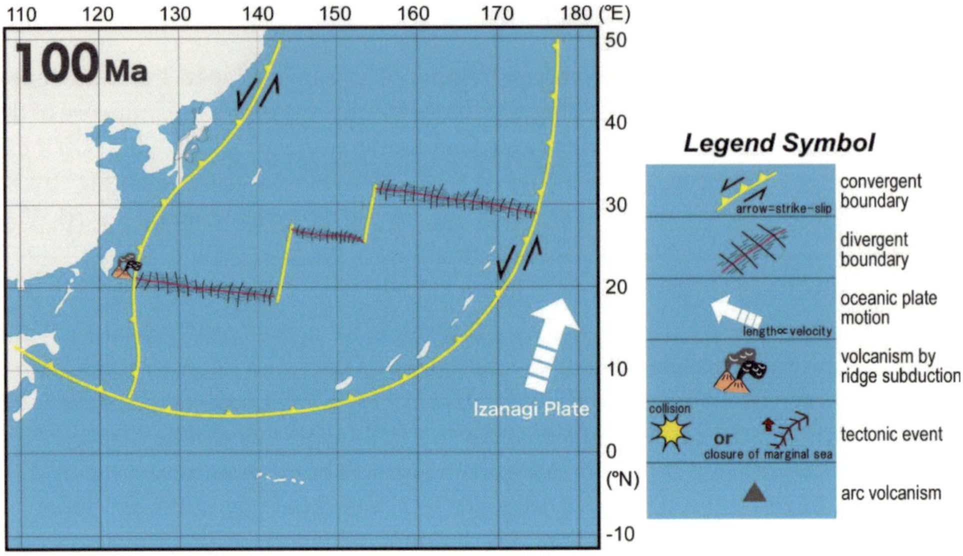

Figure 3. Reconstruction at 100 Ma.

available at OPERA: Osaka Prefecture University Education and Research Archives (http://hdl.handle.net/10466/14970).

3.1. 100 to 80 Ma

At the beginning of the Late Cretaceous, the present northwestern Pacific region was governed by the Izanagi Plate, which was characterized by quite rapid northward movement (**Figure 3**; [1]). This was promptly replaced by the westerly moving Pacific Plate (**Figure 4**). Their configuration between epochs was interpolated based on the reconstruction cartoons of Ref. [1] and calibrated by their linear velocity tables.

The eastern margin of Eurasia experienced voluminous igneous activity in the Late Cretaceous that has a clear trend toward younger activity as we move northeast. We value Kinoshita's [7] original hypothesis, assuming the effect of Kula-Pacific ridge subduction. However, the Kula-Pacific ridge's pathway in his theory does not fit with Engebretson's [1] plate reconstruction. In its place, we assume a spreading center within the newly introduced marginal sea plate. In our model, the ridge moves at a pace concordant with the time-progressive plutonism on the continental margin. It is plausible that rapid northerly migration of the Izanagi-Pacific ridge caused intense short-lived igneous activities in the Late Cretaceous along the eastern boundary of the marginal sea plate (including a major portion of the present northeast Japan forearc) (**Figure 4**), which Kinoshita [7] originally interpreted as the Farallon-Izanagi ridge effect.

Based on a detailed stratigraphic study of the Late Jurassic to Early Cretaceous Sorachi Group in central Hokkaido, Takashima et al. [20] proposed a tropical to subtropical origin of the terranes in the northern territory based on frequent occurrences of oolitic limestone and other paleontological evidence. Although the precise paleoposition of the landmass on the marginal sea plate is

http://dx.doi.org/10.5772/67358

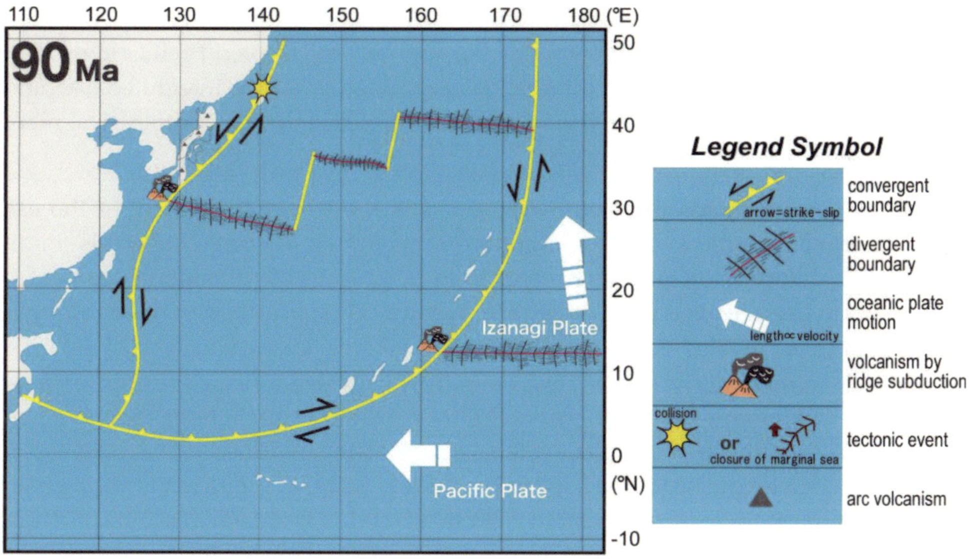

Figure 4. Reconstruction at 90 Ma.

still difficult to estimate, we regard their findings as an essential constraint in the reconstruction model. The eastern boundary of the marginal sea plate was also a site of intensive wrench deformation caused by highly oblique subduction of the Izanagi Plate. Sasaki's [15] structural analysis provides us with a clue to estimate an effective range for this wrench deformation event.

Paleomagnetic analyses by Tamaki and Itoh [10] clearly showed that the muddy sediments of the Late Cretaceous Yezo Group obtained from the Oyubari area in central Hokkaido have a geographical affinity to the mother Eurasian continent. We, hence, assume that some of the crustal blocks composing the present Japanese archipelago were settled on the margin of the mother continent for a long time and placed them upon the western convergent boundary of the newly introduced marginal sea plate.

As shown in **Figure 4**, an intraoceanic remnant arc collided against the continental margin at ca. 90 Ma after Ref. [6]. Their theory leads us to imagine an intensive deformation event in the Late Cretaceous on the Eurasian margin, facing the northwesterly moving marginal sea plate.

Collision of a remnant arc mentioned above inevitably provoked strong deformation of the forearc region. A geochemical expedition by Tamaki et al. [9] clarified that the indented rim of the continent was intensively deformed, which is recorded as an anomalously rapid burial of the Yezo forearc basin and a prompt exhumation accompanied by the buildup of overturned structures. This is a significant tectonic episode in the Late Cretaceous, and we anticipate that more geological findings can be related with this event.

This period was marked not only by a collision of landmasses upon the northern portion of the continental margin but also by a growing influence of the spreading center of the marginal sea plate. Wallis et al. [8] obtained Lu-Hf ages centered at 89–88 Ma for eclogite samples of the Sanbagawa metamorphic rocks in southwest Japan, identical to Ar-Ar phengite ages of the same unit. These ages require an extremely rapid exhumation of the metamorphosed terrane on the forearc. Their original concept was that the approaching Izanagi-Pacific ridge triggered uplift of the accretionary prism. Our alternative interpretation for the integrated thermochronological results is that the northerly moving spreading center of the hypothetical marginal sea plate caused the island arc to pop up.

Conventional wisdom holds that southwest Japan was a stable component of the eastern Eurasian margin before the Miocene backarc opening of the Japan Sea. Under this condition, the fact that its deformation structure seems to indicate multiple directional shifts of shear through the Late Cretaceous is quite enigmatic. Tokiwa [21] explicitly described the sequence of stress-strain state wobbles within the island arc based on a detailed analysis of deformation structure of an accretionary prism on its forearc. He suggested a flickering shift of sinistral-to-dextral (90–85 Ma) and dextral-to-sinistral (75–70 Ma) deformation on the same convergent margin. The northerly moving spreading center of the assumed marginal sea plate in our model acts as a toggle switch to solve the paradox, namely, segments of the convergent margin on its opposite sides show shear in reverse directions as shown in our reconstructions (e.g., **Figure 4**).

3.2. 80 to 60 Ma

Throughout the final stage of the Cretaceous and early Paleogene, the Pacific Plate was characterized by moderate northwesterly convergence as depicted in our reconstructions at 80 and 70 Ma (**Figures 5** and **6**). Its motion vector was calibrated based on the linear velocity tables of Ref. [1]. It is well known that a regional unconformity, the so-called "K/T Gap", developed in this period around the northwestern Pacific [14]. We have not yet found a drastic change in plate motion or continental collision related to this significant event.

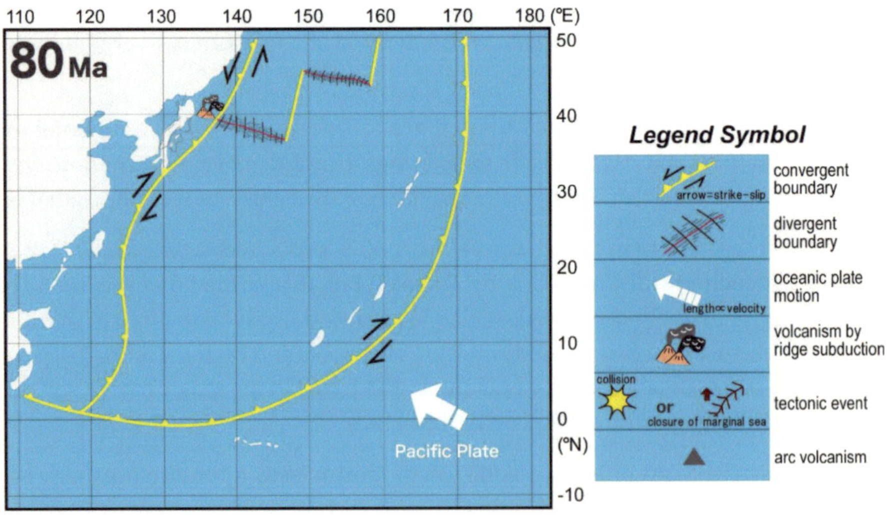

Figure 5. Reconstruction at 80 Ma.

http://dx.doi.org/10.5772/67358

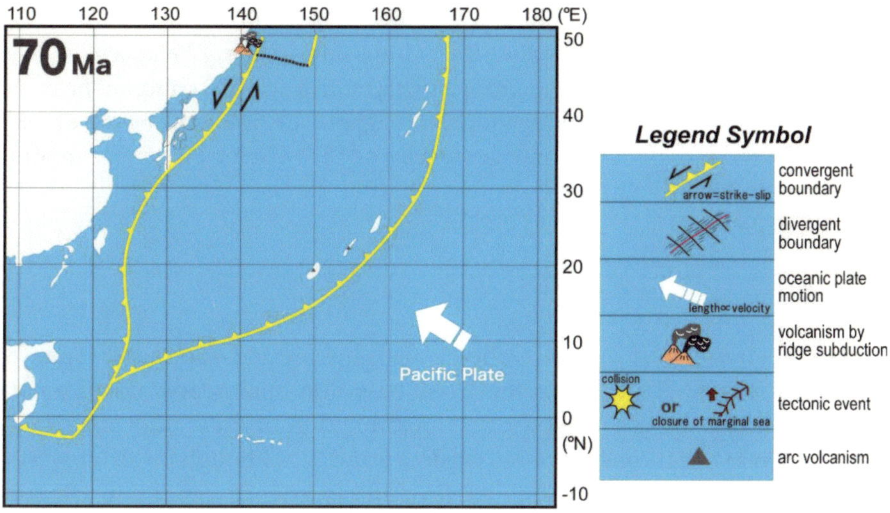

Figure 6. Reconstruction at 70 Ma.

Westerly motion of the Pacific Plate in this period may have provoked right-lateral wrench deformation of the forearc of northeast Japan, as depicted by the complex rotation sequence of a large pluton in Kamaishi revealed in the paleomagnetic research of Ref. [16]. It seems that temporal changes in the sense of the oblique convergence of the oceanic plates affected the whole forearc as presented by a paleomagnetic study of subsurface core samples by Itoh and Tsuru [17]. To evaluate rotational motions related to the forearc wrenching proposed in previous structural studies, they did paleomagnetic measurements of core samples obtained from the MITI Sanriku-oki, an exploration borehole on the forearc shelf of northeast Japan. The Eocene and Late Cretaceous cores were successfully oriented by correlating the bedding planes on the core surface with sidewall imaging, or by northing calibration, referring to the mean declination of secondary normal magnetization. Their primary directions of magnetization were corrected for tectonic tilting. Comparison with contemporaneous reference directions indicated an 87° counterclockwise rotation between the late Campanian and Eocene, and a 42° clockwise rotation thereafter. Although this sequence of rotational motions is similar to that revealed for the pluton on land (cf. [16]), sinistral wrenching seems to have lingered longer on the shelf, which may be related to a time lag in fault activity. As for the western convergent margin, Tokiwa [21] suggested the second shift of shear deformation in southwest Japan from dextral to sinistral (75–70 Ma), which is related to the demise of the spreading center on the marginal sea plate (**Figure 6**).

The low latitude origin of the Late Cretaceous Yezo Group was clearly indicated by a paleomagnetic study by Tamaki et al. [22] executed in south central Hokkaido. Their results infer that a certain part of the northeastern Japan arc may have been deposited on a remote forearc on the eastern side of the marginal sea plate. Considerable northward transportation of a part of Sakhalin was shown by a paleomagnetic study in Ref. [2]. The paleolatitudes of that study fit with the convergent trajectory of the Pacific Plate. Another paleomagnetic study in

Ref. [3], however, suggested that Sakhalin contains autochthonous blocks. Careful paleomagnetic measurements executed on the latest Cretaceous to early Paleocene volcaniclastic rocks by Otofuji et al. [23] revealed a tropical origin of a forearc component of northeast Japan. Their paleolatitudes, deduced from paleomagnetic inclinations, however, are significantly lower than those expected from the trajectory of the Pacific Plate. Our model implies that the studied landmass was on the southern part of the boundary between the marginal sea and Pacific Plates.

3.3. 60 to 40 Ma

No reliable paleostress trend for the earliest Paleogene around southwest Japan has been published. We discussed the trend in this book based on microscopic fracture network in Late Cretaceous plutonic rocks (Chapter 4). A possible shift of shear direction from sinistral to dextral is shown in our reconstructions (**Figures 6** and **7**). Although the driving force of the transient stress state has not been identified, the early Paleogene is a period when an asthenospheric injection provoked crustal thinning and formation of intracontinental basins and deformation of the convergent margin [24]. Such an event may have triggered the change in stress state.

From 55 Ma to the present, a reliable, exquisite plate reconstruction model including the southern part of our modeled area was submitted by Hall [25]. According to his compilation, some of today's components of Southeast Asia were confined adjacent to the approaching India, and spreading of the marginal sea in the Pacific region remained stagnant at around 55 Ma. The Philippine Sea Plate had begun to emerge in the equatorial Pacific region at 50 Ma (**Figure 8**). In the southern Pacific, Australia began to migrate northward at around 45 Ma, generating new convergent margins fringing around the Sundaland. The spreading centers in the Celebes Sea and Philippine Sea were active under a backarc setting.

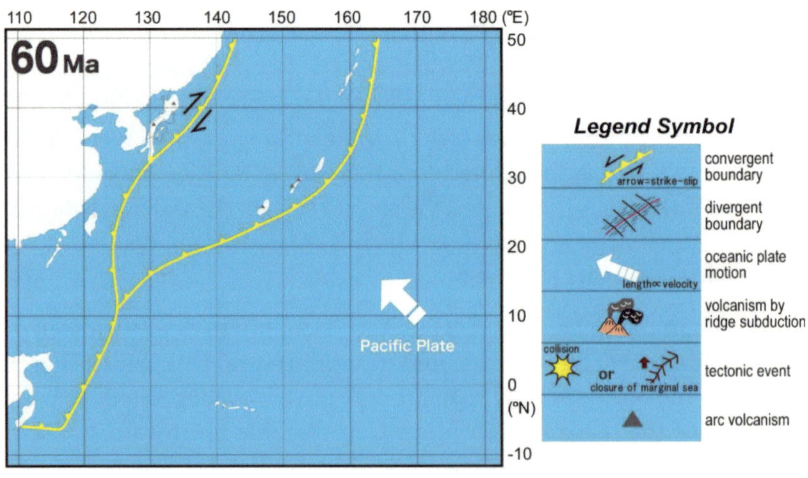

Figure 7. Reconstruction at 60 Ma.

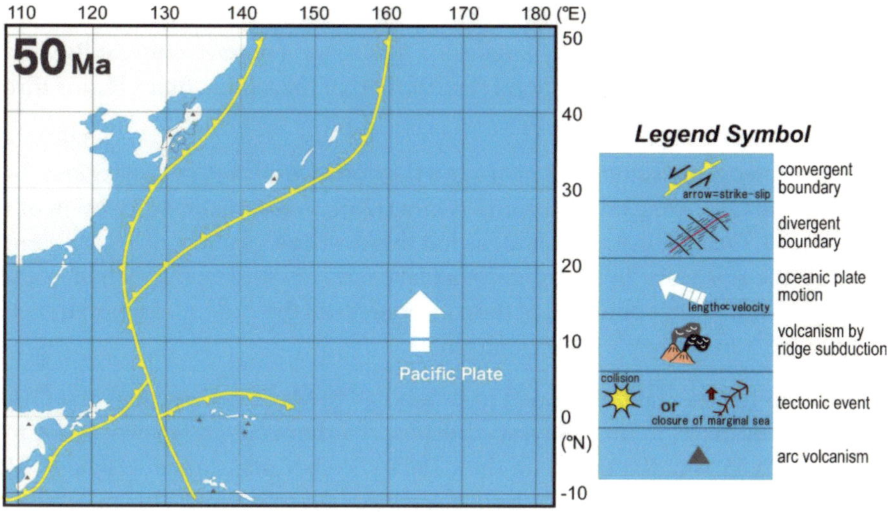

Figure 8. Reconstruction at 50 Ma.

3.4. 40 to 20 Ma

As expressed by the famous bight between the Emperor Sea Mount Chain and the Hawaiian Islands, the Pacific Plate shifted its motion counterclockwise around 45–40 Ma (**Figure 9**). Since then, the Pacific Plate has been steadily moving westward. Its linear velocity in each epoch was calibrated based on data tables in Ref. [1].

After the Middle Eocene development of fluvial to estuarine basins buried by coal-bearing clastics, central Hokkaido and the forearc area of northeast Japan became a site of intermittent subsidence [26]. Our model relates this phenomenon to shifting of the Pacific Plate motion

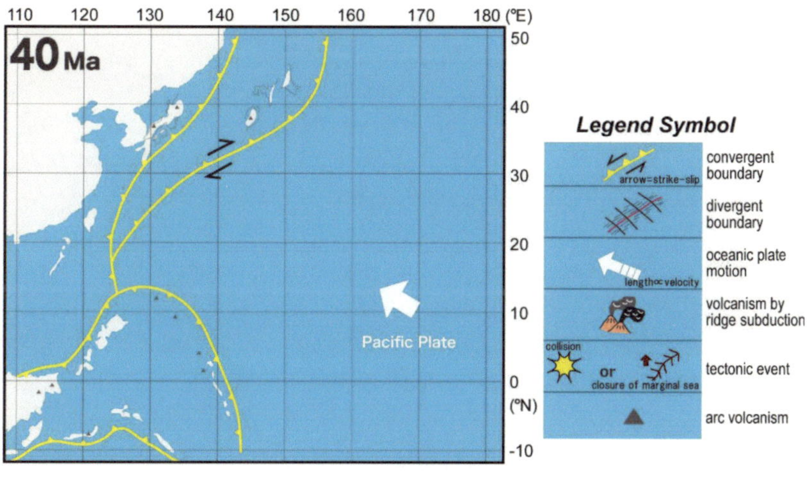

Figure 9. Reconstruction at 40 Ma.

and prevailing dextral shear on the eastern border of the marginal sea plate. Numerical models constructed in a geophysical study suggested that several transcurrent faults with dextral slips can restore the actual spatiotemporal distribution of the sedimentary basins throughout the Paleogene [18].

The most noted Paleogene tectonic event around the northwestern Pacific margin is the formation of the regional Oligocene unconformity (Ounc; [26]), a systematic description of which is given in Chapter 1. Our reconstruction regarded this event as an effect of strong compressive stress raised by the closure of the marginal sea plate and subsequent collision of island arcs at ca. 30 Ma (**Figure 10**). At that point, all of the components of the present-day arc-trench system had arrived at the northwestern Pacific margin.

The proto-Philippine Sea Plate continued to expand at 40 Ma [25]. The azimuth of its spreading center had been fixed in an eastwest direction. To the west, India was about to collide against Eurasia. At 35 Ma, divergence in the Celebes Sea became stagnant, whereas the east-west trending ridge in the proto-Philippine Sea continued to spread. The early phase of the formation of the Philippine Sea Plate had come to an end by ca. 30 Ma. The marginal sea was surrounded by subduction zones.

In Hall's [25, 27] reconstruction, the Philippine Sea Plate began to rotate clockwise around the beginning of the Miocene (24 Ma) accompanied with strike-slip (sinistral) motion in northern New Guinea. We can see an embryotic spreading center along the northeastern margin of the plate. It seems, however, that many ambiguous points remain with regard to the kinematic model of the rotation event. Recently, Kimura et al. [28] advocated that the Philippine Sea Plate swiftly rotated clockwise nearly simultaneous with the clockwise motion of southwest Japan driven by the Miocene backarc opening of the Japan Sea. On the other hand, Yamazaki et al. [29] stated that the main rotation phase of the Philippine Sea Plate took place before 25 Ma based on newly obtained paleomagnetic data from the northwestern part of the plate. The authors refrain from giving final judgment on such a chaotic condition because further investigation based on firm geochronological information is needed.

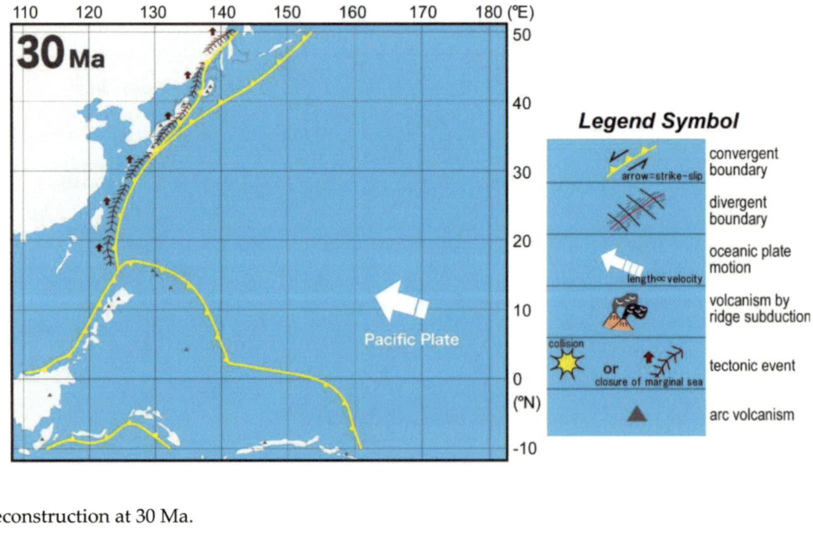

Figure 10. Reconstruction at 30 Ma.

http://dx.doi.org/10.5772/67358

3.5. 20 Ma to present

An interpretation of reflection seismic data covering the offshore forearc region of northeast Japan [17] depicted a series of transcurrent faults cutting slantwise through the arc. Based on a conspicuous trend of offshore geomagnetic anomalies, the distribution of Paleogene arc volcanism and a provenance study of the coeval detritus within forearc basins, Itoh and Tsuru [17] restored the arc-arc junction, and advocated more than 200 km of dextral displacement on the fault zone since the Oligocene period. Its ongoing status is delineated in the reconstruction at 20 Ma (**Figure 11**).

In addition to prevailing dextral motion on an arc-bisecting fault zone, numerical modeling by Kusumoto et al. [18] clarified that an eastwest compression and reverse fault motion must have emerged to restore the configuration of the elongate basin of the Middle Miocene Kawabata Formation in central Hokkaido. This may be linked to the initiation of the collision between the Kurile and northeast Japan arcs.

Although some ambiguous points remain, energetic paleomagnetic studies by Otofuji and his colleagues (e.g., [30]) confirmed that the rifting and backarc opening in the Japan Sea occurred in a relatively short period not later than 15 Ma. The spatiotemporal distribution of marine sediments on the Japanese backarc supports their working hypothesis. A paleomagnetic study in the eastern part of southwest Japan revealed that arc bending had occurred after the Japan Sea backarc opening. Based on a reconstruction of terrane arrangement, Itoh [31] argued that the collision of the proto-Izu arc had precipitated the bending event, which is a significant constraint on the coeval position of the eastern margin of the Philippine Sea Plate. Geological evidence in the South Fossa Magna tectonic zone suggests multiple collisions of landmasses on the Izu-Bonin arc, namely the Kushigatayama, Misaka, Tanzawa, and present Izu Peninsula [32]. It seems that the eastern margin of the Philippine Sea Plate has been anchored in front of the easternmost part of southwest Japan since 15 Ma.

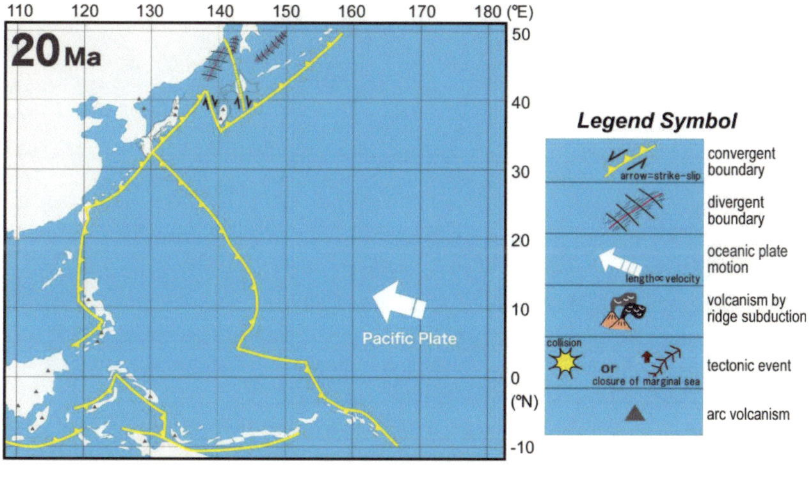

Figure 11. Reconstruction at 20 Ma.

Regardless of the theories described in the previous section we adopt, clockwise rotation of the Philippine Sea Plate and spreading of the Shikoku Basin were about to terminate by 15 Ma. Within the Philippine Sea Plate, extension of the Sula Spur was still active, forming the North Banda Sea by ca. 10 Ma. To the west of our mapped area, the Java Trench subduction zone began to roll back, causing extension of the Sundaland Margin [25]. At around 5 Ma, the South Banda Sea was being formed, whereas the Molucca Sea subduction was almost complete and the Halmahera and Sangihe arcs were about to collide. We can see volcanism as a precursor to rifting on the Mariana arc (**Figure 12**). A marine geological survey and geomagnetic anomaly modeling revealed the latest rifting event in the Okinawa Trough since the end of Pliocene [33]. Our model incorporates its configuration. Spreading on the Mariana arc still continues now, and the basin floor is widening [25].

Around southwest Japan, the most remarkable event in the second half of this interval is a strong contraction on its backarc side and the formation of a regional unconformity. Itoh and Nagasaki [34], based on interpretation of seismic reflection data, clarified that an eastwest folded zone on land continues onto the whole backarc shelf. Although the driving force of the tectonic episode has not been fully elucidated, some hypothesize that revitalized subduction of the Philippine Sea Plate is responsible for the regional arc deformation (e.g., [28]). If this is the case, the motion history of the oceanic plate should be reexamined in light of deformation trends on the convergent margin.

Based on submarine geomorphology and geology around the collision front of the Izu Peninsula, Nakamura et al. [35] advocated that a counterclockwise shift of the converging direction of the Philippine Sea Plate occurred around 2–1 Ma. Their finding has a great significance in the field of active tectonics because west-northwestward motion of the oceanic

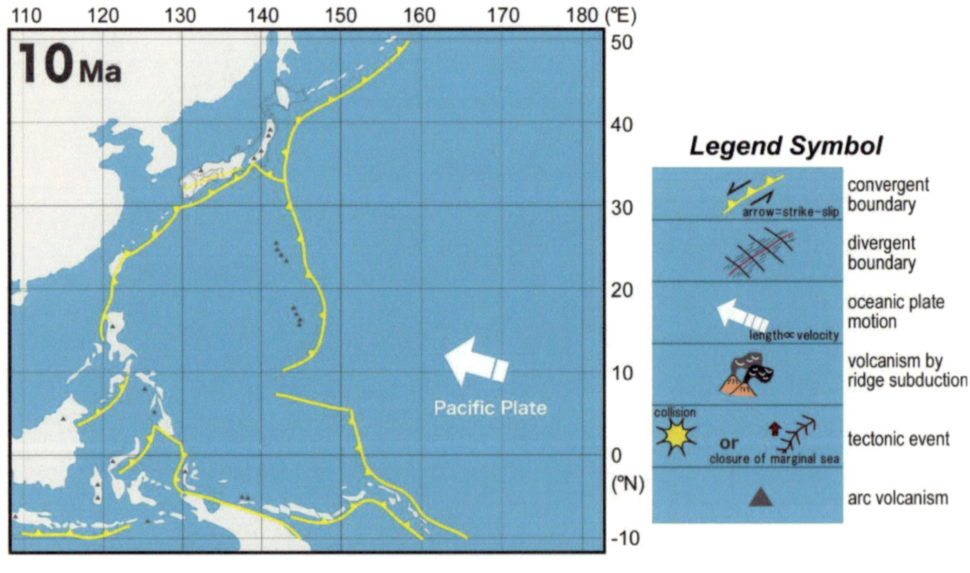

Figure 12. Reconstruction at 10 Ma.

plate enhances right-lateral slips on the arc-bisecting Median Tectonic Line and development of intra-arc basins controlled by activity on numerous secondary faults in southwest Japan. A temporal change in the motion of the Philippine Sea Plate also caused extensive wrench deformation of southwest Japan. Based on an integrated analysis of seismic data, geological and geomorphological evidence, Itoh et al. [36] clarified that the island arc is a continental sliver put between regional transcurrent faults with active dextral slips. A closer look at the spatiotemporal deformation trends in southwest Japan [37] shows a complex pattern of contraction and extension features, which are comprehensively understood as the effect of a transient converging azimuth and migration of the Euler pole of the Philippine Sea Plate. Further investigation of tectonic episodes on the circum-Pacific convergent margins will pave the way for improved plate reconstruction.

4. Conclusions

Our thoroughgoing examination has indicated the spatiotemporal positions of diverse tectonic events around the northwestern Pacific. The evolution of the Pacific margin is viewed as a history of migration/amalgamation of allochthonous blocks onto the subduction zone. In order to reconcile paradoxical discrepancies in the docking process of genetically grouped arc components like the "forearc basin", the authors introduced a marginal sea plate with a spreading center that was alive in the Cretaceous. Oblique subduction of the ridge caused specific migratory igneous activity along the rim of the overriding plates, together with flips of shearing direction. Arc-trench systems on the eastern and western sides of the marginal sea plate developed following different timelines and were eventually mixed up during the Oligocene closure of the plate. This phenomenon is now recognized as a coincident regional clinounconformity on the Eurasian margin. Since the demise of the hypothetical plate, the tectonic regime of the northwestern Pacific margin has been controlled by the growth of the Philippine Sea Plate. Accurate restoration of its rotational history and modes of convergence should be pursued through further quantitative research on the geodynamics of active circum-Pacific margins.

As for the present result, a live show is available on the platform of Dagik Earth (three-dimensional digital globe), which is a project by the visualization group of the earth science hub of Kyoto University (http://earth.dagik.org). Anyone can access our plate model (http://dagik.org/menu/land/Dagik_NWPacific_100my/e) as animated cartoons by means of a web browser.

Acknowledgements

Discussions with Miho Sasaoka were a great help in improving an early version of the illustrations of the plate reconstruction model. We are grateful to Akinori Saito for his earnest support during the course of preparation of the contents for the Dagik Earth.

Author details

Yasuto Itoh[1]*, Osamu Takano[2] and Reishi Takashima[3]

*Address all correspondence to: yasutokov@yahoo.co.jp

1 Graduate School of Science, Osaka Prefecture University, Osaka, Japan

2 Technical Division, Japan Petroleum Exploration Co. Ltd., Tokyo, Japan

3 The Center for Academic Resources and Archives (Tohoku University Museum), Tohoku University, Sendai, Japan

References

[1] Engebretson DC, Cox A, Gordon RC. Relative motions between oceanic and continental plates in the Pacific Basin. Geological Society of America Special Paper. 1985; 206: 1-59.

[2] Bazhenov ML, Zharov AE, Levashova NM, Kodama K, Bragin NY, Fedorov PI, Bragina LG, Lyapunov SM. Paleomagnetism of a Late Cretaceous island arc complex from South Sakhalin, East Asia: convergent boundaries far away from the Asian continental margin? Journal of Geophysical Research. 2001; 106: 19193-19205.

[3] Weaver R, Roberts AP, Flecker R, Macdonald DIM, Fot'yanova LM. Geodynamic implications of paleomagnetic data from Tertiary sediments in Sakhalin, Russia (NW Pacific). Journal of Geophysical Research. 2003; 108: 2066. doi:10.1029/2001JB001226

[4] Natal'in B. History and modes of Mesozoic accretion in Southeastern Russia. The Island Arc. 1993; 2: 15-34.

[5] Ueda H, Kawamura M, Niida K. Accretion and tectonic erosion processes revealed by the mode of occurrence and geochemistry of greenstones in the Cretaceous accretionary complexes of the Idonnappu Zone, southern central Hokkaido, Japan. The Island Arc. 2000; 9: 237-257.

[6] Ueda H, Miyashita S. Tectonic accretion of a subducted intraoceanic remnant arc in Cretaceous Hokkaido, Japan, and implications for evolution of the Pacific northwest. The Island Arc. 2005; 14: 582-598.

[7] Kinoshita O. Migration of igneous activities related to ridge subduction in Southwest Japan and the East Asian continental margin from the Mesozoic to the Paleogene. Tectonophysics. 1995; 245: 25-35.

[8] Wallis SR, Anczkiewicz R, Endo S, Aoya M, Platt JP, Thirlwall M, Hirata T. Plate movements, ductile deformation and geochronology of the Sanbagawa belt, SW Japan: tectonic significance of 89–88 Ma Lu-Hf eclogite ages. Journal of Metamorphic Geology. 2009; 27: 93-105.

[9] Tamaki M, Tsuchida K, Itoh Y. Geochemical modeling of sedimentary rocks in the central Hokkaido, Japan: episodic deformation and subsequent confined basin-formation along the eastern Eurasian margin since the Cretaceous. Journal of Asian Earth Sciences. 2009; 34: 198-208.

[10] Tamaki M, Itoh Y. Tectonic implications of paleomagnetic data from upper Cretaceous sediments in the Oyubari area, central Hokkaido, Japan. Island Arc. 2008; 17: 270-284.

[11] Gilder S, Courtillot V. Timing of the North-South China collision from new middle to late Mesozoic paleomagnetic data from the North China block. Journal of Geophysical Research. 1997; 102: 17713-17727.

[12] Ueda H. Accretion and exhumation structures formed by deeply subducted seamounts in the Kamuikotan high-pressure/temperature zone, Hokkaido, Japan. Tectonics. 2005; 24: TC2007. doi:10.1029/2004TC00169

[13] Ando H, Tomosugi T. Unconformity between the Upper Maastrichtian and Upper Paleocene in the Hakobuchi Formation, north Hokkaido, Japan: a major time gap within the Yezo forearc basin sediments. Cretaceous Research. 2005; 26: 85-95.

[14] Ando H. Stratigraphic correlation of Upper Cretaceous to Paleocene forearc basin sediments in Northeast Japan: cyclic sedimentation and basin evolution. Journal of Asian Earth Sciences. 2003; 21: 921-935.

[15] Sasaki M. Early Cretaceous sinistral shearing and associated folding in the South Kitakami Belt, northeast Japan. The Island Arc. 2003; 12: 92-109.

[16] Itoh Y, Takano O, Kusumoto S, Tamaki M. Mechanism of long-standing Cenozoic basin formation in central Hokkaido: an integrated basin study on an oblique convergent margin. Progress in Earth and Planetary Science. 2014; 1: 1-14. doi:10.1186/2197-4284-1-6

[17] Itoh Y, Tsuru T. A model of late Cenozoic transcurrent motion and deformation in the forearc of northeast Japan: constraints from geophysical studies. Physics of the Earth and Planetary Interiors. 2006; 156: 117-129.

[18] Kusumoto S, Itoh Y, Takano O, Tamaki M. Numerical modeling of sedimentary basin formation at the termination of lateral faults in a tectonic region where fault propagation has occurred. In: Itoh Y, editor. Mechanism of Sedimentary Basin Formation—Multidisciplinary Approach on Active Plate Margins. Rijeka, Croatia: InTech; 2013. pp. 273-304. doi:10.5772/56558

[19] Kawamura M, Yasuda N, Watanabe T, Fanning M, Terada T. Composition and provenance of the Jurassic quartzofeldspathic sandstones of the Oshima accretionary belt, SW Hokkaido, Japan. Memoirs of the Geological Society of Japan. 2000; 57: 63-72.

[20] Takashima R, Nishi H, Yoshida T. Late Jurassic-Early Cretaceous intra-arc sedimentation and volcanism linked to plate motion change in northern Japan. Geological Magazine. 2006; 143: 753-770. doi:10.1017/S001675680600255X

[21] Tokiwa T. Timing of dextral oblique subduction along the eastern margin of the Asian continent in the Late Cretaceous: evidence from the accretionary complex of the Shimanto Belt in the Kii Peninsula, Southwest Japan. Island Arc. 2009; 18: 306-319.

[22] Tamaki M, Oshimbe S, Itoh Y. A large latitudinal displacement of a part of Cretaceous forearc basin in Hokkaido, Japan: paleomagnetism of the Yezo Supergroup in the Urakawa area. Journal of the Geological Society of Japan. 2008; 114: 207-217.

[23] Otofuji Y, Sato K, Iba N, Matsuda T. Cenozoic northward translation of the Kitakami massif in northeast Japan: paleomagnetic evidence. Earth and Planetary Science Letters. 1997; 153: 119-132.

[24] Itoh Y, Uno K, Arato H. Seismic evidence of divergent rifting and subsequent deformation in the southern Japan Sea, and a Cenozoic tectonic synthesis of the eastern Eurasian margin. Journal of Asian Earth Sciences. 2006; 27: 933-942.

[25] Hall R. Cenozoic geological and plate tectonic evolution of SE Asia and the SW Pacific: computer-based reconstructions, model and animations. Journal of Asian Earth Sciences. 2002; 20: 353-431.

[26] Takano O, Itoh Y, Kusumoto S. Variation in forearc basin configuration and basin-filling depositional systems as a function of trench slope break development and strike-slip movement: examples from the Cenozoic Ishikari—Sanriku-Oki and Tokai-Oki—Kumano-Nada forearc basins, Japan. In: Itoh Y, editor. Mechanism of Sedimentary Basin Formation—Multidisciplinary Approach on Active Plate Margins. Rijeka, Croatia: InTech; 2013. pp. 3-25. doi:10.5772/56751

[27] Hall R. Late Jurassic-Cenozoic reconstructions of the Indonesian region and the Indian Ocean. Tectonophysics. 2012; 570-571: 1-41.

[28] Kimura G, Hashimoto Y, Kitamura Y, Yamaguchi A, Koge H. Middle Miocene swift migration of the TTT triple junction and rapid crustal growth in southwest Japan: a review. Tectonics. 2014; 33: 1219-1238. doi:10.1002/2014TC003531

[29] Yamazaki T, Takahashi M, Iryu Y, Sato T, Oda M, Takayanagi H, Chiyonobu S, Nishimura A, Nakazawa T, Ooka T. Philippine Sea Plate motion since the Eocene estimated from paleomagnetism of seafloor drill cores and gravity cores. Earth Planets Space. 2010; 62: 495-502.

[30] Otofuji Y, Hayashida A, Torii M. When was the Japan Sea opened?: paleomagnetic evidence from Southwest Japan. In: Nasu N, Uyeda S, Kushiro I, Kobayashi K, Kagami H, editors. Formation of Active Ocean Margins. Tokyo: Terra Publishing Co.; 1985. pp. 551-566.

[31] Itoh Y. Differential rotation of the eastern part of southwest Japan inferred from paleomagnetism of Cretaceous and Neogene rocks. Journal of Geophysical Research. 1988; 93: 3401-3411.

[32] Amano K. The South Fossa Magna as a multi-collision zone. Gekkan Chikyu (Earth Monthly). 1986; 8: 581-585.

[33] Sibuet JC, Letouzey J, Barrier F, Charvet J, Foucher JP, Hilde TWC, Kimura M, Chiao LY, Marsset B, Muller C, Stephan JF. Back arc extension in the Okinawa Trough. Journal of Geophysical Research. 1987; 92: 14041-14063.

[34] Itoh Y, Nagasaki Y. Crustal shortening of Southwest Japan in the Late Miocene. The Island Arc. 1996; 5: 337-353.

[35] Nakamura K, Renard V, Angelier J, Azema J, Bourgois J, Deplus C, Fujioka K, Hamano Y, Huchon P, Kinoshita H, Labaume P, Ogawa Y, Seno T, Takeuchi A, Tanahashi M, Uchiyama A, Vigneresse JL. Oblique and near collision subduction, Sagami and Suruga Troughs—preliminary results of the French-Japanese 1984 Kaiko cruise, Leg 2. Earth and Planetary Science Letters. 1987; 83: 229-242.

[36] Itoh Y, Tsutsumi H, Yamamoto H, Arato H. Active right-lateral strike-slip fault zone along the southern margin of the Japan Sea. Tectonophysics. 2002; 351: 301-314.

[37] Itoh Y. Gunchu Formation—An Indicator of Active Tectonics on an Oblique Convergent Margin. Germany: LAP LAMBERT Academic Publishing; 2015. 76 p.